D1091220

Microwave Materials and Fabrication Techniques

Third Edition

For a listing of recent titles in the *Artech House Microwave Library*,
turn to the back of this book.

Microwave Materials and Fabrication Techniques

Third Edition

Thomas S. Laverghetta

Artech House
Boston • London
www.artechhouse.com

Library of Congress Cataloging-in-Publication Data
Laverghetta, Thomas S.
Microwave materials and fabrication techniques / Thomas S. Laverghetta.—3rd ed.
p. cm. — (Artech House microwave library)
Includes bibliographical references and index.
ISBN 1-58053-064-8 (alk. paper)
1. Microwave circuits—Materials. 2. Microwave circuits—Design and construction.
I. Title. II. Series.

TK7876 .L378 2000 00-040623
621.381'33—dc21 CIP

British Library Cataloguing in Publication Data
Laverghetta, Thomas S.
Microwave materials and fabrication techniques.—3rd ed.
(Artech House microwave library)
1. Microwave circuits
I. Title
621.3'8132

ISBN 1-58053-064-8

Cover design by Igor Valdman

685 Canton Street
Norwood, MA 02062

International Standard Book Number: 1-58053-064-8
Library of Congress Catalog Card Number: 00-040623

10 9 8 7 6 5 4 3 2 1

Contents

Preface

The second edition of this book came out in 1991 and was an update of the original text. At that time, it was the most up-to-date version that could be placed on the market. Well, it is now 2000 and the idea of being up-to-date has gone by the wayside, which is why a third edition has been put together. This book is an attempt to present the most recent information possible to those who need to know about microwave materials and how to fabricate the circuits that either they or someone else designed. There will probably come a day when this book does not accomplish that task, but for now it has the information designers need.

Chapter 1 is a rewrite of the second edition and basically introduces the reader to the ins and outs of microwaves, why some of the terms are used, and why dimensions are important in microwaves.

Chapter 2 deals with microwave materials (laminates and substrates). There probably is no greater area of change in the microwave industry over the past 10 years than that of microwave materials. The shift from military to commercial applications has brought forth a wide variety of new materials that will fit these new requirements, and these materials are covered in this chapter.

The only change made to Chapter 3 is its introduction, which was rewritten to further emphasize the importance of a microwave designer being familiar with not only the electrical properties of a design but also the metals and other materials that are a part of the complete design.

Chapter 4 covers microwave artwork and is divided into initial layout, final layout, and photo process. The chapter covers the schematic/layout programs that are now available. Designers can use the computer to design the circuit, lay out the artwork for the circuit, and have artwork produced that is accurate and repeatable. Some of the new programs that are currently available are presented.

Chapter 5 has been expanded. In the second edition, this chapter was called Etching Techniques. This third edition expands this section to cover both Etching and Plating Techniques. The addition of plating processes and discussions of plated-through-holes add a very valuable section to this book.

Chapter 6 covers bonding techniques. This chapter is basically the same as the second edition with the solders and epoxies updated to reflect what is actually available today.

Chapter 7 is on microwave packaging and covers stripline, microstrip, and suspended substrate packaging. It also includes a small section from the second edition on microwave connectors. The final section in this chapter covers electromagnetic interference (EMI) considerations, which is a very important part of microwave packaging.

The intention of this book is the same as the previous two editions—to educate microwave circuit designers in materials and fabrication techniques. If designers are familiar with the techniques discussed in each chapter, they will certainly avoid many problems and have a design that will perform as well on the bench as it does on paper or on a computer screen—and maybe even better.

Thomas S. Laverghetta
Auburn, Indiana
July 2000

Acknowledgments

When you keep putting together more and more editions of a book, you rely on many people to provide information and keep you up to date so that you can put accurate material in the book. There have been many people who have helped me put this edition together.

I would like to say a very large thank you to Phil Johnson of GIL Technologies. He made a super effort to come to Fort Wayne and sit down to answer questions that I had about certain areas of the industry. I know how busy he is and I really appreciate his help, even when I called to ask more questions. There was always an answer . . . thanks, Phil.

I would also like to thank four people who provided data sheets for microwave material so that I could have the most recent sheets in the Appendix. These people are Curt Zimmerman of GIL Technologies, Donna Weber of Arlon, Barry Beumer of Taconic, and Karen Krider of Rogers. All it took was a call and a request and the information was sent to me within a couple of days. Thank you all for your efficiency. I would also like to thank Dane Collins of Applied Wave Research for his help while I was working on the artwork chapter. The information he supplied was of great help.

There is always a special young lady—Joanne Olinger—whom I thank in every one of my books. Once again, she was there when I needed to have some drawings done and did an excellent job. There were not a lot of new drawings for this edition, but every one of them was clear and accurate and something that I was proud to submit to Artech House.

Finally, I would once again like to thank my family for being patient during the process of putting together this new edition. It is a much more difficult task to do a following edition than it is to do an entirely new book. My family knows this and cut me the extra slack that I needed to finish the job. There would be no life at all without them because they keep me going and keep me in check many times.

1

Introduction

When beginning a microwave circuit design, there are many factors that must be taken into consideration. Frequency of operation, tolerable losses, gain requirements, and dc voltages are common parameters that are routinely considered when the design process begins. These are excellent parameters to consider, but they are not the only ones that should be investigated. Right from the start, the designer should also consider what type of material is to be used for the circuit, how connections are going to be made to that circuit, what type of package is best for that circuit, and how the circuit board is to be attached to that package. The last set of parameters presented are probably the ones that are always placed at the very end of the design process. They are usually put in the category of "oh, by the way." (The circuits that we are referring to here are the microstrip and stripline circuits used for many microwave applications. The assumption throughout this book is the use of stripline or microstrip. If there are cases for coaxial or waveguide applications, they will be specifically referred to as such.)

The first area that we referred to was the type of material to be used in the new microwave circuit that we will design. Chapter 2 covers in detail the types of materials currently available to a designer. It also introduces the reader to the appropriate terminology necessary to communicate with manufacturers and other technical people.

Many times, you will see a circuit board referred to as a printed circuit board, regardless of the frequency of operation. This is fine if we are talking about a circuit that is used in a commercial AM/FM radio, television, or a computer-controlled toy. It is, however, the wrong designation for a microwave circuit board. The conventional circuit board, as referred to earlier, is designed to support components and provide a path for current to flow between these components. The actual properties of the materials, other than mechanical and environmental, are generally of no use to a designer. The electrical properties can vary all over the place and there will be no adverse effect on the circuit operation. This is not the case for the microwave circuit board. The board itself is an integral part of the entire microwave circuit. As such, it is imperative that the designer investigate many different types of material and look at all of the electrical, thermal, and mechanical properties of that material before using it for the final circuit.

The parameter that is the most recognized and most used for microwave materials is the dielectric constant. This parameter tells a designer what frequency range the material should be used over. (Chapter 2 will discuss more details on this and other material parameters.) By recognizing certain materials by their dielectric constants, a designer can narrow the choice of materials very rapidly and speed up the decision process.

To see how important this one parameter is to having proper operation of the finished microwave circuit, consider the following example. Suppose that we have an area of ½ × ½ in a stripline circuit on which we are to place a 50Ω line. The length of the line is to be one full wavelength at 2 GHz. Let us say that there are two laminates available for use in this circuit. One has a dielectric constant of 2.1 and the other is 10.2. (This is not a good situation because you should not have to be limited to only one or two types of laminate material. However, for this example, we will limit ourselves to only two.)

The first step in determining the proper laminate from the two that we have made available is to find out just how long one full wavelength is for each of the materials. To do this, we will use a very common formula for wavelength:

$$\lambda = \frac{c}{f\sqrt{\varepsilon}}$$

When this is done, the resulting values are:

$$\lambda_{\varepsilon=2.1} = 4.07''$$

and

$$\lambda_{\varepsilon=10.2}\ 1.85''$$

From the calculations shown above, it can be seen that the lengths are significantly different. The full wavelength for the 2.1 material is quite a bit longer than the 10.2 material. Since we have only a ½″ × ½″area to work in, it would seem that we need to use the shorter line in order to get the whole transmission line on the circuit board. The best way to check this theory is to draw each of the lines on a scaled ½ × ½ square to see how they fit. A very important part of the drawing is the width of the 50Ω line on each of the materials being considered. For the 10.2 material, the width, W, of the 50Ω line is 0.010″, and for the 2.1 material it is 0.051″ for the same thickness of material. It can be seen that these dimensions will have a large effect on just how much of the full wavelength we can put into the allotted space. This is further complicated by the fact that the spacing between lines is critical when meandering transmission lines on a circuit board. A spacing of one ground plane spacing (the thickness of the two boards that make up the stripline package) between transmission lines must be maintained to eliminate any possibility of crosstalk between lines.

Figure 1.1 shows the 10.2 material with its 0.010″ transmission line meandered across the required ½ × ½ area. It can be seen that the full wavelength is accommodated with more than adequate spacing between transmission lines. Similarly, Figure 1.2 shows the 2.1 material in the same area with the proper spacing between lines. You will note that there is less than one-half wavelength (0.45λ) in the same area in which we had a complete wavelength for the 10.2 material. So it can be seen that there is a large

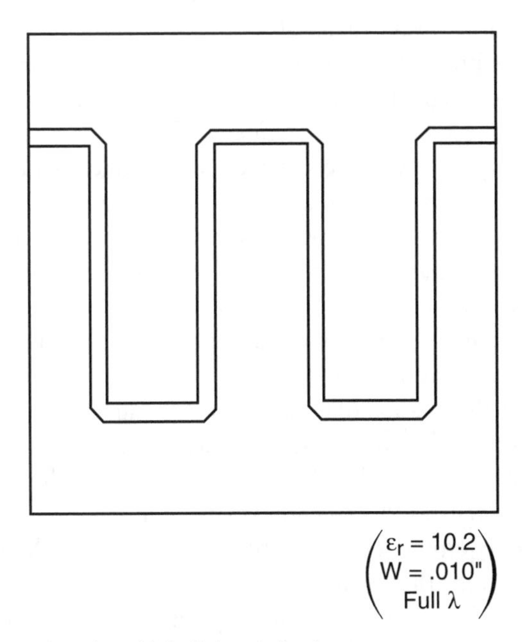

Figure 1.1 Full wavelength on high-dielectric laminate.

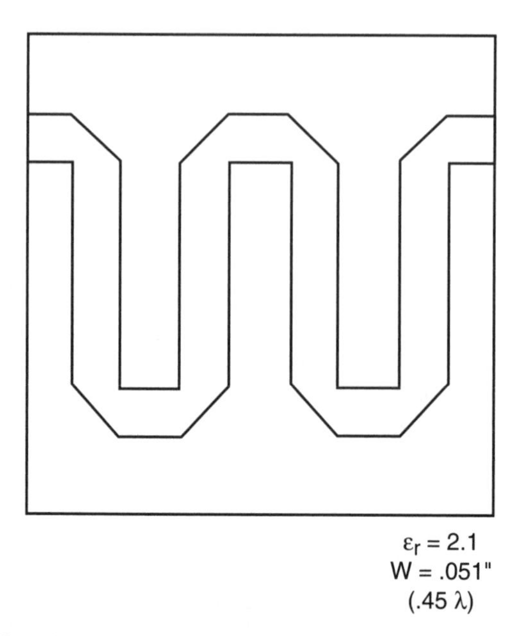

Figure 1.2 Fraction of a wavelength on high-dielectric laminate.

impact on the design of a microwave circuit even when only considering one parameter of the microwave material.

Other areas previously mentioned are the attachment of components to the microwave circuit board and placing the circuit into its proper finished package. The first area, attachment of components, can be accomplished in a variety of ways. The specific type of attachment will depend on the application you have. You may have surface mount technology (SMT) or monolithic microwave integrated circuits (MMIC) to consider in your particular application. As such, you need to consider them in a different light than the typical microstrip or stripline configuration. Also, such factors as mechanical and environmental conditions that the finished circuit will be exposed to need to be addressed.

Soldering is the method of attachment that usually comes to mind when discussing the topic. This, of course, is a very common method of attaching components to a circuit board and can be seen in every radio, television, camcorder, or VCR available. In fact, the solder joints are many times suspect when this type of equipment does not perform as expected. Many times, there is a "cold" solder joint that is causing the problem. This is an improper joint that works for a while but becomes intermittent and causes problems.

The most common type of solder, and the one that comes to mind first, is the family of tin-lead (SnPb) solders. The combination of these two elements may be such that there is a 50/50 combination, a 60/40 split favoring either material, or other combinations of the two. These combinations result in various melting temperatures and will be explored further later in the book.

Elements other than tin and lead are also used for solders used for microwave circuit and component attachment. Such metals as indium (In), silver (Ag), gold (Au), cadmium (Cd), and zinc (Zn) are commonly used for a variety of reasons, depending on each individual metallic interface. The metallic interface is when you have a component or substrate with a certain metal on it (copper, gold, etc.) and a base plate for a case with another metallic contact (aluminum, gold, etc.). The interface between the metals must be compatible. Many times, the interface is the solder you will use and it must be an efficient interface that will allow the circuit to operate properly and not cause any additional problems.

An interface problem that can occur is when a tin-lead solder is used on gold. If, for example, you use 60/40 tin-lead solder (60% tin, 40% lead) on a gold metallized substrate, there will be a metallic interface between the tin and the gold that results in a brittle joint that will not be a reliable connection. To eliminate this problem, the designer has two options. First, use a solder with no tin in it. Without the tin, the problem does not exist. Indium solders are used for many of these applications. One possibility is a 50/50 InPb solder or other combinations that use indium and eliminate the use of tin.

The use of indium solders may not be possible for your particular application. If this is the case, a solder can be used that has a very low content of silver. An example of this is a ratio of 62.5% tin, 36.1% lead, and 1.4% silver. The silver will inhibit the interaction, or leaching, of the gold and tin and help to eliminate the brittle joint problem. This example should serve to illustrate the need to know the metallic content of all the elements that make up a microwave circuit. The content of the substrate metallization, the case used to house the circuit, and the solder used to attach components or substrates must all be considered before an operational circuit can be realized.

We briefly discussed the use of solder to attach components or substrates, but epoxies can be used to accomplish the same purpose. Use of epoxy is a precise method for attaching components without having them exposed to the high temperatures of a soldering process. The epoxy is cured at a specific temperature and for a specified time duration. Some typical temperatures and times are 80°C (170°F) for 1½ hours; 120°C (248°F) for 1 hour; or 150°C (302°F) for a two-part epoxy (resin and hardener). For the one-part epoxy, the temperature and time would be in the range of 150°C (302°F) for 2 hours. More details on one- and two-part epoxies will be presented in Chapter 6.

To wrap up this introduction, we will look at how the circuit is to be packaged. Chapter 7 will discuss this concept in detail, but certain areas will be addressed here.

There are many critical areas to consider when dealing with packaging of microwave circuits. One is the transition from a microwave circuit board to the coaxial connector that interfaces with the outside world. Figure 1.3 shows such a transition. It can be seen from Figure 1.3(a) that the substrate is connected to a case with the tab for the connector soldered to the transmission line on the substrate. (This connection could also be made using an epoxy.) This is probably the type of connection that always comes to mind when a connection from substrate to connector needs to be made. When there are no severe temperature constraints put on the circuit to be built, this is an excellent method for making the connection.

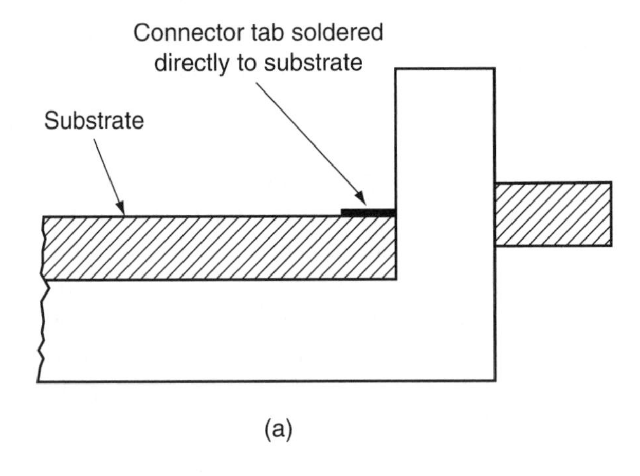

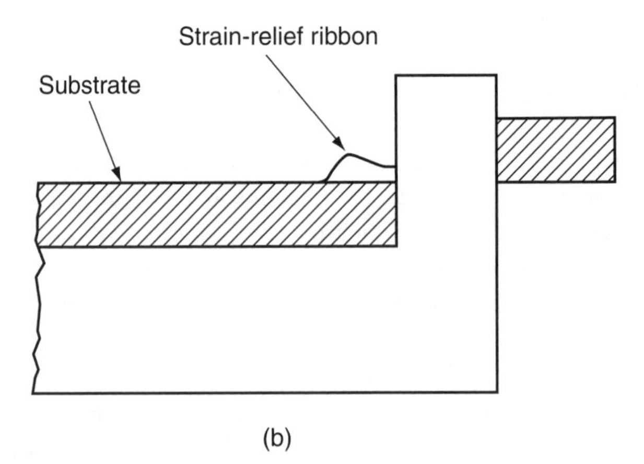

Figure 1.3 Substrate-to-connector transitions.

However, if the circuit is to operate over a wide temperature range, very serious problems can occur. Every material has a coefficient of thermal expansion (CTE) that tells an individual how much a material will change its dimensions for every degree of temperature change. If you have the same materials for all applications, there is no problem because the materials all change by the same amount. This, however, is a very rare occurrence. Usually, you have a variety of materials present and they all move at different rates. This results in stresses being put on certain areas—usually the connection between the substrate and the connector—of the circuit. The result is either a cracked connection or a completely broken connection.

To minimize the effects of temperature on connections for microwave circuits, certain things can be done. You may want to investigate all of the materials you will be using and try to choose materials that have CTEs that are very close to one another. You can also change the connection method, as shown in Figure 1.3(b). This has a mechanical strain relief placed between the connector and the substrate that allows the materials to move and take the stress off the solder or epoxy joint. In theory, this is a good idea. However, this strain relief should be looked at very carefully. If the relief is too large, the results will be as shown in Figure 1.4.

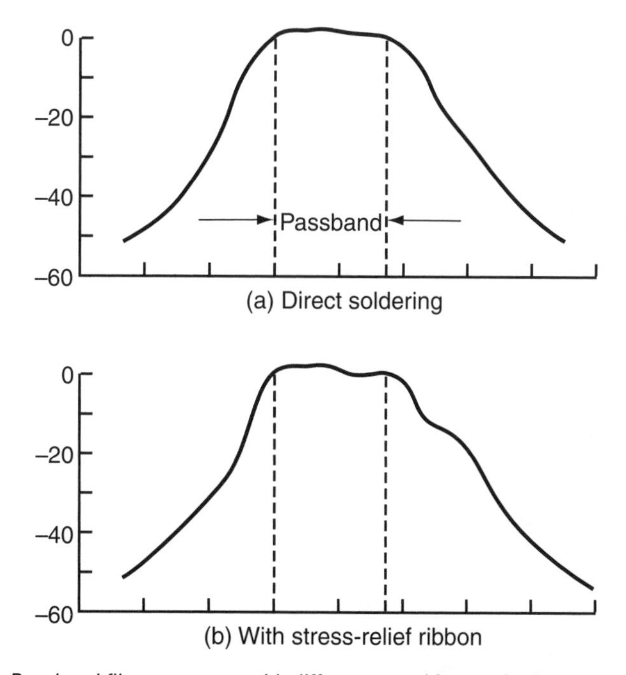

Figure 1.4 Passband filter response with different transition methods.

In Figure 1.4(a), a filter response is shown that has the connector tabs soldered directly to the circuit. It can be seen that there is a good filter response shown. In Figure 1.4(b), the stress is relieved by a loop in the connection and the response is not good at all. This is not to say that this type of relief is always bad. If the relief is small enough, there will be very little effect on the circuit, and the mechanical forces are taken care of.

Thus, it can be seen that there is much more that goes into a microwave circuit design than to simply etch a few lines on a circuit board, drill some holes, and put it in a case. The microwave circuit that works well and for a long time is one that the designer has spent sufficient time in choosing the right circuit board material, properly attaching the components to the circuit, properly attaching the material to a case, properly connecting the circuit to a connector, and choosing the right case design for the circuit. A circuit that looks nice on your desk does very little to make a system work or get you praise from your boss. The circuit that works well is the ultimate goal and takes time and effort to achieve.

2

Laminates and Substrates

2.1 Introduction

If you were designing a microwave circuit in the 1960s, your choice of material was pretty straightforward. You could use 3M K6098, 3M K6098, or 3M K6098. This was a woven Teflon fiberglass material that was 0.062″ thick, had a dielectric constant of 2.55, and had 1 oz copper on both sides. (The terms used will be discussed in detail later in this chapter.) The whole trick at that time was to be sure that the K6098 would exhibit close to the properties you needed over your frequency range. This was usually a significant challenge because of a wide variety of specifications that would not necessarily fit the material.

Today is a much different story for the microwave designer. There are in excess of 30 different types of materials available from which to choose. The challenge now is not whether the material exhibits the correct properties but which material will do the best job for you. Should your material be woven, nonwoven, ceramic-based, polyester, a thermoset, a PTFE/ceramic, or a pure Teflon® material? (PTFE stands for polytetrafluorethylene, which is the chemical terminology for Teflon.) To make the proper choice, many factors must be considered. This chapter will define terms and consider the factors that go into choosing the *correct* material for your particular application.

Before proceeding further, it is necessary to make a distinction between two terms that will be referred to many times and are the title of this chapter. These terms are "laminate" and "substrate." Usually, when the word "laminate" is used, it is in reference to "soft" substrates such as Teflon (PTFE)-based materials. These materials may have fiberglass or ceramic reinforcement in them or may be a composite material. Many are actually constructed with a lamination process, similar to the process used to laminate Social Security cards or driver's licenses. They are referred to as "soft" substrates to categorize them for microwave applications. This does not mean that they are so soft that they will not hold their shape. They are simply referred to as soft to indicate their flexibility, which allows them to be easily sheared, drilled, or milled. Conversely, when a material is specifically called a substrate, it is usually a "hard" material. Many pure ceramic materials are hard substrates and will be referred to simply as a substrate.

2.2 Microwave Material Terms

In order to make an intelligent choice of a material for a particular application, it is necessary to understand the terminology used for microwave materials. This knowledge will help you interpret the material data sheet and make an informed choice. The terms to be presented are dielectric constant, anisotropy, dissipation factor, dielectric thickness, copper weight, coefficient of thermal expansion, and peel strength. These are common terms that are widely used and will characterize a microwave material sufficiently so that you can make the required decision.

2.2.1 Dielectric Constant

In order to understand the term *dielectric constant,* you must first know what a dielectric is. If you look up the term "dielectric" in the dictionary, it says that a dielectric is an insulator. That is, it will not pass current in a circuit. This is an accurate definition of a dielectric if we are applying it to power line insulators or the material in capacitors. It is not, however, accurate when working with microwave materials. To understand what the difference is between a dielectric in a capacitor and a dielectric as it applies to microwave materials, consider the following example.

A large pipe is suspended by wires with a target attached to the wall at the end of the pipe. If the pipe is empty, we can throw a ball through the pipe and hit the target with little effort. If we fill the pipe with some loosely packed feathers, it will take a bit more energy to get the ball through the feathers, but you probably will still hit the target. If we now fill the pipe with water and take some license to assume that the water will not fall out, it will be very difficult to get the ball through the water to the target.

In the example, the ball remained the same, the pipe remained the same, and the target remained the same. The only parameter that changed was the medium through which the ball was to be thrown. In the second and third cases (feathers and water), we slowed down the ball and impeded its path to the target. That is, we *obstructed* the ball by changing the medium through which it has to travel. That is exactly what a microwave dielectric does. It obstructs the microwave energy as it propagates through the material. The more dense the material, the slower the energy moves through it. Thus, we can say that a dielectric in microwaves is:

> a material that creates an environment causing microwave energy to be reduced from what it would be in free space.

Now that we have a good idea of what a dielectric is, we can look at the more common parameter that is found on microwave material data sheets, the dielectric constant. This term (sometimes referred to simply as DK or permitivity) goes along with the definition we just derived. The dielectric constant describes a material's density relative to air. With the dielectric constant of air being 1, all other materials have dielectric constants greater than 1. As an example, pure Teflon has a dielectric constant of 2.10 (sometimes listed as 2.08 in certain tables). This means that the velocity of a microwave signal through this material is decreased to 69% of what it would be in air. This is achieved by taking the square root of the dielectric constant (2.1), which is 1.4491, and taking the reciprocal of that number, which is 0.69 or 69%. Similarly, a dielectric constant of 3.5 decreases the velocity of the signal to 53.4% (1/1.87) of that in air, and a dielectric constant of 10.2 decreases the velocity to 31.3% (1/3.1937) of the velocity in air. It should be apparent at this point that the higher the dielectric constant, the slower the velocity of the microwave energy propagating through it because a higher dielectric constant means a more dense material.

As a rule of thumb, you should use a low dielectric constant material when operating at the high end of the microwave spectrum. This will result in much more reasonable lengths for the transmission lines on the circuit board. Similarly, materials with a higher dielectric constant should be used at the lower end of the microwave spectrum. These guidelines are for size considerations. At low frequencies, the wavelengths get considerably longer, as we saw in the example in Chapter 1. By using a higher dielectric constant material, the lengths are easier to handle and are put on a circuit board to maintain an architecture that will be small and conform to the RF and microwave size constraints.

The statements made above concerning the use of certain dielectric constants are general. In reality, there is a tremendous overlap of dielectric constants as a function of frequency. This is illustrated in Figure 2.1. From this figure, you can see that there is usually more than one dielectric constant that can be used in a particular frequency range. If, for example, your frequency

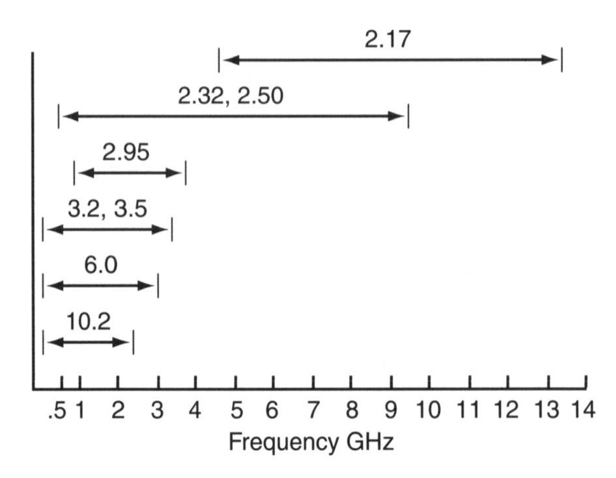

Figure 2.1 Dielectric constant chart.

range was from 2–3 GHz, the figure shows that any one of four dielectric constants could be used: 2.95, 3.2, 3.5, or 6.0. Thus, it can be seen why more than one parameter is used to choose a material for a particular design, with dielectric constant being but one of these parameters.

2.2.2 Anisotropy

A term relating to dielectric constant that is not always taken into consideration is *anisotropy.* To understand anisotropy, you must first understand that the dielectric constant that has been presented thus far is the dielectric constant in the X-Y plane. This is designated as ε_{xy}. There is also a dielectric constant in the Z-plane of the material that is designated as ε_z. The term anisotropy is a comparison of these two dielectric constants. That is, anisotropy is $\varepsilon_{xy}/\varepsilon_z$. Some typical values of anisotropy are shown in Table 2.1.

Table 2.1 shows a limited number of materials as an illustration. The ultimate goal for anisotropy is to have it attain a value of 1.0, where the dielec-

Table 2.1
Anisotropy

Material	$\varepsilon_{xy}/\varepsilon_z$
Woven, $\varepsilon = 2.17$	1.090
Woven, $\varepsilon = 2.45$	1.160
Nonwoven, $\varepsilon = 2.20$	1.025
Nonwoven, $\varepsilon = 2.33$	1.040

tric constant is the same for both the X-Y plane and the Z-plane. This, however, is not a practical state and, thus, the anisotropy should be as close to 1.0 as possible. It should be noted that for the materials shown in the table (which are all a combination of Teflon and fiberglass materials), the dielectric constants are much closer for the nonwoven materials than for the woven materials. Also, it is much closer to 1.0 for the lower dielectric constant material. This can be explained by the fact that at lower dielectric constants, there is less fiberglass added to the base Teflon material and, thus, there is a much more uniform structure. Also, the nonwoven structure is more uniform than the typical woven structure, accounting for the better anisotropy for those materials. An excellent example of a low anisotropy is a pure Teflon material. This is one of the most uniform materials that can be used for microwave applications and has an isotropy that is basically 1.0 because of the uniform structure.

Anisotropy is an important parameter to consider when you have a critical application where the dielectric constant must be controlled in all planes.

2.2.3 Dissipation Factor

Every circuit ever developed, whether in microwaves, at low frequencies, or when considering dc or ac power, has certain losses that are inherent in it. The same is true of microwave materials. There is a loss within the material that is called the *dissipation factor* (many times referred to simply as DF).

If you hear the term "dissipation," you usually think of something being lost. When you hear heat dissipation, you can visualize heat pouring from a radiator or furnace. When you hear on the news that the early morning fog will dissipate by 10 A.M., you picture it gone by that time. The term "dissipate" in microwaves has a similar meaning. As we previously said, dissipation is a loss. In microwaves, dissipation is a loss of energy in the form of heat. Thus, in the case of microwave laminates or substrates, we want the material to store energy within the structure rather than lose it in any way. The storing of the energy will result in a low loss path for the microwave energy. (The terminology of storing energy refers to the energy remaining in the circuit rather than being sent off as heat, causing the material to exhibit a high loss.)

The dissipation factor can also be referred to as Tan δ. This implies, rightly so, that the dissipation factor is actually a ratio. It is the ratio of energy dissipated to energy stored:

$$\text{Tan } \delta = \text{Energy Dissipated/Energy Stored}$$

This is illustrated in Figure 2.2, which shows the loss angle of a material. This loss angle (δ) is determined by the relationship between the loss term and the conductivity of the material, as shown in Figure 2.2.

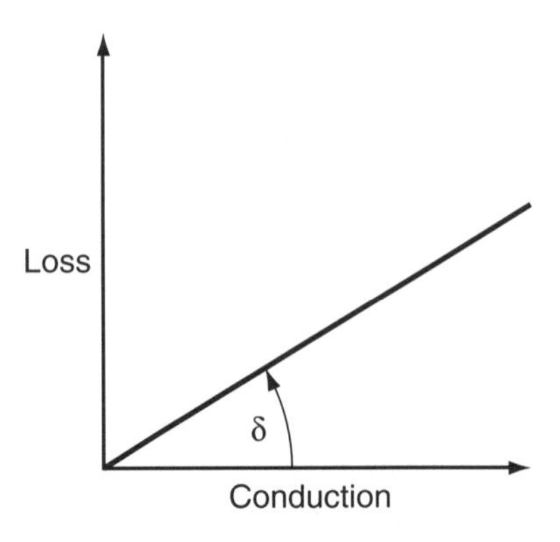

Figure 2.2 Dissipation factor.

If we consider the energy dissipated as a loss, or resistance, and the energy stored to be a measure of conductivity of the material, we now have a relationship that says:

$$\text{Dissipation Factor} = \text{Loss(Resistance)/Conductivity}$$

To have an efficient microwave material, you need a low value of dissipation factor. From the relationships presented thus far, you can see that this is possible when the resistance is low or the conductivity is high. Some representative materials and their associated dissipation factors are shown in Table 2.2. (These numbers are valid at 10 GHz.)

Typical dissipation factors for specific materials will be presented as we progress through the chapter and introduce materials for discussion. For each value of dissipation factor, there should be an associated frequency with

Table 2.2
Dissipation Factors

Material	Dissipation Factor (Tan δ)
Glass Epoxy	0.0180
PTFE/Glass	0.0018
Alumina	<0.0001
Polyethylene	0.0002
Mica	0.0026
Silicon Rubber	0.0032
Ethyl Alcohol	0.0620

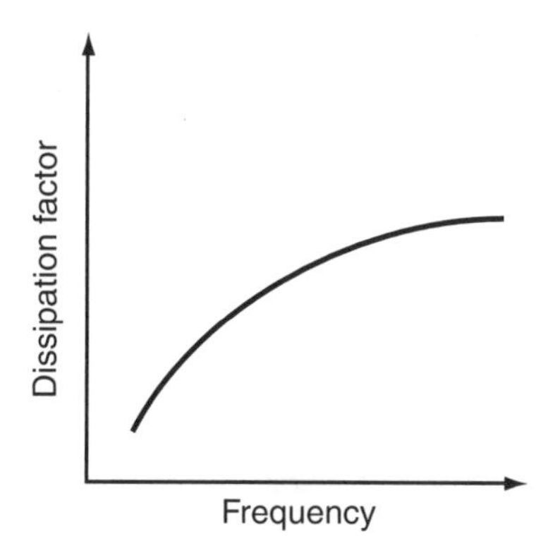

Figure 2.3 Dissipation factor versus frequency.

it. This is because, unlike the dielectric constant that is basically not a function of frequency, the dissipation is a factor of frequency. As the frequency of operation increases, so does the dissipation factor. The change in dissipation factor is not a linear function. That is, it is not a simple straight line with an increase in frequency. It is more of an exponential function as shown in Figure 2.3, which is a representative curve that you should look for when consulting a data sheet for a microwave material. Thus, it can be seen that the dissipation factor is a very important parameter that should be considered in all cases, with the particular frequency of operation also noted.

2.2.4 Dielectric Thickness

In Chapter 1, we discussed putting a transmission line on a piece of microwave material and compared two different dielectric constants to see how the line would fit in a specified space. We said that for each dielectric constant, there was a specific width of transmission line "for the same thickness of material." The last part of that statement is in quotes because it is a very important part of choosing a particular microwave material for an application. This parameter, called the *dielectric thickness*, is usually referred to as "b" for a ground plane spacing in a microwave circuit. This is a designation that will be investigated further in the next few paragraphs.

To fully appreciate the importance of dielectric thickness, you must be fully aware of exactly what thickness we are talking about. When you look at a laminate or substrate, you will see a piece of material with copper on both sides and some dark brown or white material in the center; gold completely surrounding a base material; or a white base material with gold on two sides. These combinations are shown in Figure 2.4.

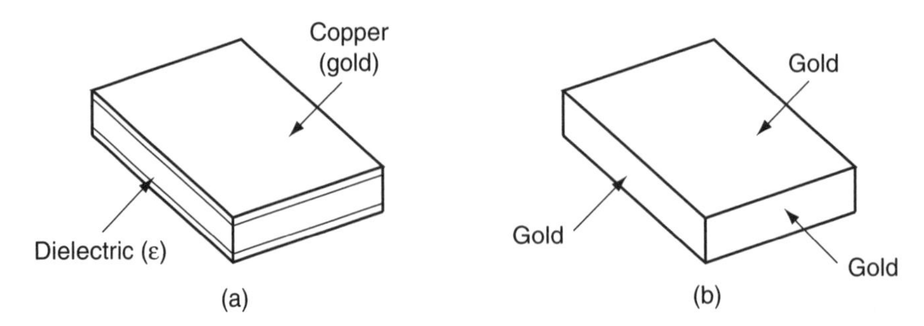

Figure 2.4 Microwave laminates and substrates.

Figure 2.4(a) shows the laminate or substrate with copper (or gold) on both sides but not on the edges. Figure 2.4(b) is a representation of a substrate with metallization on all six surfaces. (Actually, there is a combination of metals on the surfaces. These will be covered when hard substrates are presented.) If you just took these drawings and tried to determine the dielectric thickness of a material, you would be misled into thinking that the thickness was the total thickness of the material just as it appears when you hold it in your hand. The actual dielectric thickness dimension is shown in Figure 2.5. It can be seen that the dielectric thickness is the dimension of the dielectric *only*, as the name implies. Any cladding or metallization that is on the dielectric is *not* part of the thickness dimension. So when a data sheet says that a particular material is available in 0.030″ and 0.062″ sizes, it is referring to the dielectric material only. The copper or other metallization thickness is not part of this dimension. This is an additional dimension that may be taken care of later in the design procedure.

It may be easier to remember what dielectric thickness is if you remember that the thickness is the dimension of an *unclad laminate*. This is the "b" dimension shown in Figure 2.5 and referred to earlier in this chapter. As can be seen, this is the thickness of one piece of material as it is shown in the figure. There are variations to this, however. When using only one circuit board, which is microstrip, the "b" dimension is the thickness of

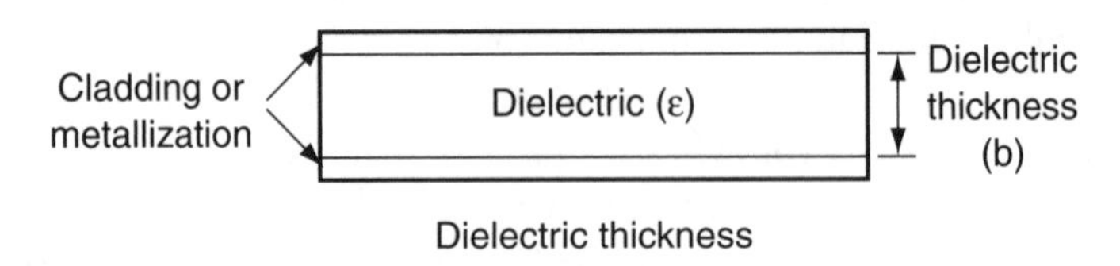

Figure 2.5 Dielectric thickness dimension.

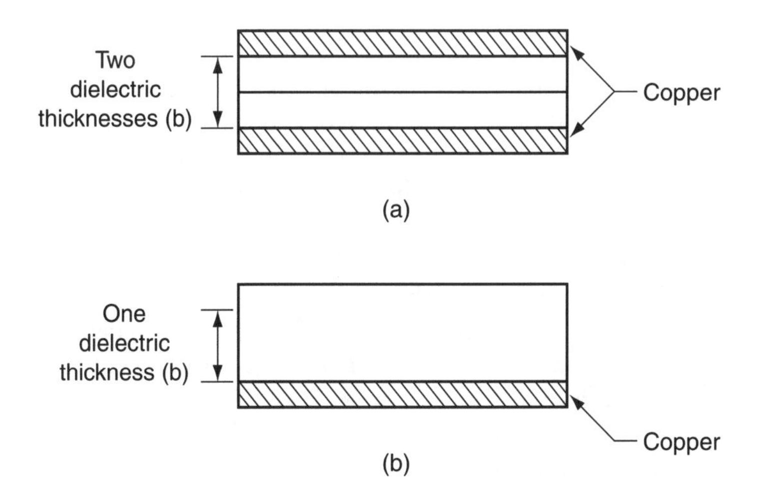

Figure 2.6 Dielectric thickness: (a) stripline and (b) microstrip.

that one piece of material. If stripline transmission lines are used, this "b" dimension is actually two pieces of microwave material stacked on top of one another. Figure 2.6 shows this relationship between dielectric thickness for both stripline and microstrip. For stripline, the "b" dimension is used in design equations to determine the width of transmission lines and the spacing between them if a coupling circuit is needed. For microstrip, the "b" dimension is also used to calculate any transmission line widths or spacings. The factor to remember in these calculations is that for microstrip, the number that is on the data sheet is the "b" dimension that you use. For stripline, it must be doubled before any calculations can be made.

There are many different thicknesses available from a variety of manufacturers. The actual thickness, however, is not the most important number to look at when choosing the thickness of a material. The tolerance of the material is significantly more important than the base thickness. It is necessary to have very close tolerance on the thickness because it is so important in determining other parameters for components, such as the characteristic impedance of the transmission line and the value of coupling that will result from two transmission lines being placed close together. Typical tolerances available are ±0.001″, ±0.0015″, or ±0.002″, which are very close tolerances. The typical epoxy circuit board that is inside a commercial radio or television does not need such tight tolerances because it is only used as a vehicle for current to get from one point to another. In a microwave circuit, the material is an integral part of the circuit and the tight tolerances are absolutely necessary.

2.2.5 Copper Weight

This usually strikes people as being a strange parameter for a microwave material. People typically wonder why they should care how much a piece of copper weighs. This is probably a very reasonable question to ask. However, this parameter is telling a designer how thick the copper is on the material, and that is a very important piece of information to have. But how do we get from a weight to a thickness? A straightforward method is used to make this conversion. The copper weight number that is on a data sheet is the weight of one square foot of copper. If you have several standard 1-square-foot sections of copper that weigh different values, the only thing that can be different is the thickness of the copper itself. Therefore, the thicker the copper, the higher the copper weight. Thus, the thickness of the copper is designated by a weight in ounces.

Standard copper weights are ½ oz, 1 oz, and 2 oz, with other special weights available upon request. The actual conversion from ounces to inches is as follows: ½ oz copper is 0.0007″ thick, 1 oz copper is 0.0014″ thick, and 2 oz copper is 0.0028″ thick. Other conversions to consider: ¼ oz copper is 0.00035″ thick and ⅓ oz copper is 0.0005″ thick.

Why is it necessary to have different weights of copper on a microwave circuit board, and where would these different weights be used? As a general rule, the less current that will be flowing in a circuit, the thinner the copper should be. If, for example, you have a very low current application, the ½ oz copper would do very well. If, on the other hand, you are designing a power amplifier that will require very high currents, the 2 oz copper would be the choice so that the actual transmission lines would not be vaporized by the high current.

Another guideline to use is the width of the transmission lines or the spacing between transmission lines. If a transmission line is very narrow or if there is a narrow gap between transmission lines, it is best to use a thin copper. For wider transmission lines and wider gaps between lines, a thicker copper is appropriate.

To understand how important it is to have the proper copper weight, consider Figure 2.7. In this figure, the top portion is an ideal cross section of the material and the copper that is on top of the material. It can be seen that the edges are very sharp and straight. This, as has been said, is an ideal representation of what an etched circuit board looks like. The bottom portion of Figure 2.7 shows the typical cross section of an etched circuit. You can see that the sides are not straight and sharp but kind of curved and undercut as a result of the etching process. This is the biggest reason for having the proper weight of copper for each application.

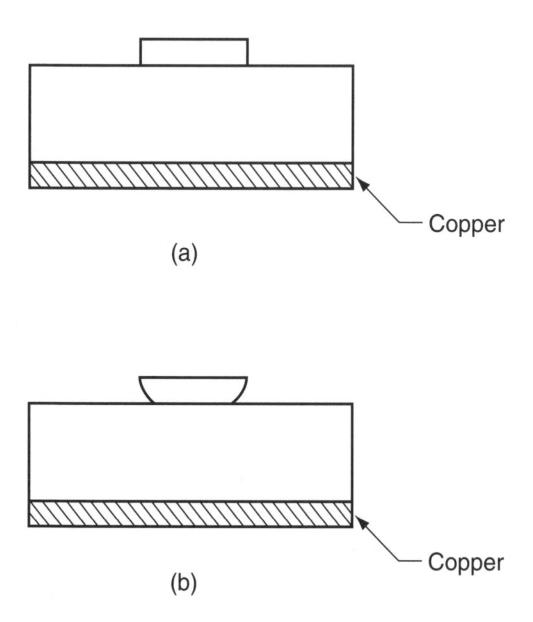

Figure 2.7 Transmission line cross sections: (a) ideal cross section and (b) actual cross section with the proper copper weight.

To expand further on this point, look at Figure 2.8. This is a comparison of 2 oz and ½ oz copper. Case I is using 2 oz copper and Case II is for ½ oz copper. In Case I, we are required to etch a very narrow gap (<0.005″) between two transmission lines. It can be seen that the lines etched on the 2 oz copper are dramatically undercut. That is, they are narrower on the underside of the line than at the top. This is because of the time required to etch the thicker copper. The etchant does not etch the top portion because of the photoresist on the surface. The lower portion is thus subjected to the etchant for a longer period of time and overetches. The ½ oz portion is virtually undisturbed. This is due to the smaller volume of copper that needs to be etched to arrive at two transmission lines separated by a narrow gap.

In Case II, there is only a single transmission line to etch, which is narrow (<0.010″). The same conditions occur when the 2 oz copper is used, as can be seen in Figure 2.8. Once again, there is undercutting that causes variations in line dimensions. You can also see that, once again, the ½ oz copper case is virtually undisturbed and the narrow line is basically intact.

One condition that must be watched in both cases is just how drastic does this undercutting become. If the circuit is etched for too long a period, the undercutting could be so severe that the line lifts completely away from

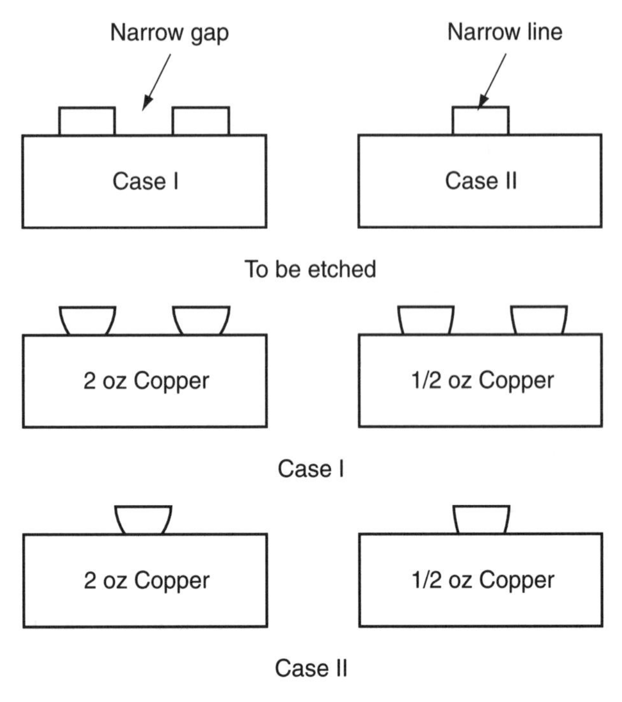

Figure 2.8 Copper weight comparison.

the laminate. This is shown in Figure 2.9. To eliminate such an occurrence, you should take great care when etching narrow lines, regardless of whether you are using thick or thin copper.

As a summary of our copper weight discussions for microwave laminates, consider the following advantages and disadvantages:

2 oz copper:

- Good copper weight for high-power applications;
- Copper will undercut on narrow gaps and lines;

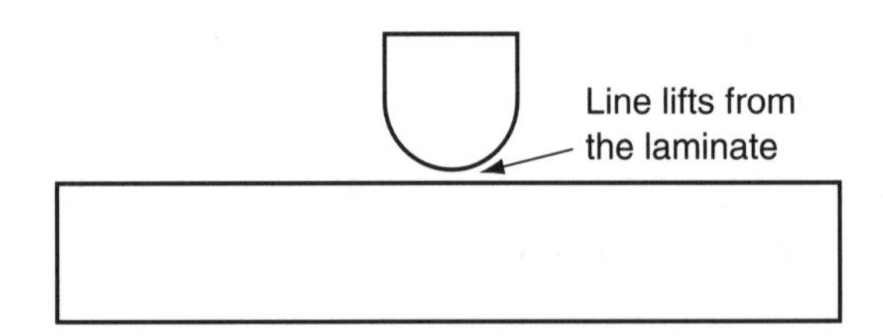

Figure 2.9 Overetching of lines.

- Narrow lines inconsistent along a long length;
- Good for general-purpose circuitry;
- A standard weight.

1 oz copper:

- A very common weight for a variety of microstrip and stripline circuits;
- Less undercut on narrow gaps and lines than 2 oz;
- Can be used for medium-power applications;
- A standard weight.

½ oz copper:

- Used for low-power applications;
- Excellent for narrow gap and line etching;
- Do not use for high-power circuits;
- A standard weight.

¼ oz copper:

- Should be used only for low-power circuits;
- Excellent for narrow gaps and lines;
- A nonstandard weight usually available in special order only.

Now that we have looked at the different weights of copper, let us see how the copper is fabricated. There are two methods of copper fabrication—rolled and electrodeposited. The standard method used for most microwave laminates is electrodeposited (ED) copper. Rolled copper is available by special request, and the data sheet usually has a statement saying the laminate can be supplied with rolled copper "for more critical electrical applications." We will cover the standard ED copper first and then the rolled.

If you were to look up the word *electrodeposition* in the dictionary, either a conventional or electronic dictionary, you would find the typical dictionary definition: it would say that electrodeposition is the process of depositing a substance on an electrode by electrolysis. Unless you are a chemist, this probably tells you no more than you knew when you first looked up the word. To clear the water a little, you could look up *electrolysis.* At this point, you would find that the dictionary says that electrolysis is the process of changing the chemical composition of a material by sending an electric current through it. If you put these two definitions together and do some thinking, you will get some idea of what ED copper is. Here is a basic definition of the term:

> *Electrodeposited copper* is a material produced by a chemical building process in which the individual copper particles are electrically joined to form the desired sheet thickness.

This can be likened to taking small pieces of wet clay and building one continuous sheet from the small pieces. If you had a certain number of these pieces of clay placed next to one another over some base surface (as ED copper is plated on a rotating drum structure and then pulled off), you would have a structure that was a certain thickness and consistency. This might be likened, for example, to plating (or depositing) ½ oz copper (0.0007″). If you needed 1 oz copper (0.0014″), you would repeat the process until the desired 1.4 mils of thickness were achieved. Similarly, you would have to add more wet clay to your already existing structure in order to achieve the required thickness. This depositing process can be very precisely controlled by controlling the time and current used in the electrodeposition process.

As previously mentioned, ED copper is the standard method for microwave laminates. Rolled copper is obtained by special request to the man-

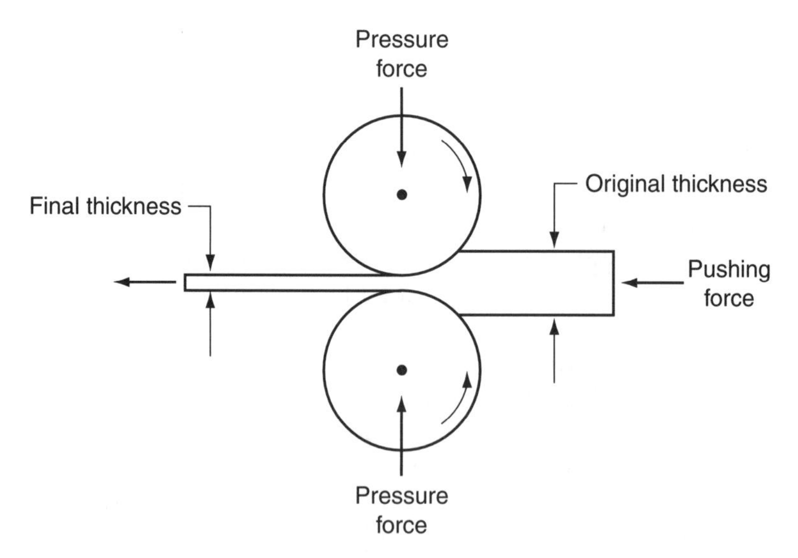

Figure 2.10 Rolled copper diagram.

ufacturers and is used for more critical applications. You will see why this is so when we complete our discussion on rolled copper and compare it with standard ED copper.

The easiest way to picture rolled copper is to compare it to putting something through an old-style wringer washer. The copper block that is used to make a copper foil for laminates is compressed in much the same manner. Figure 2.10 shows this concept. The original thickness of copper is pushed into the pair of rollers, which are being both rotated and forced together. (This vertical force depends on the final thickness required at the output.) As the copper is forced through the rollers, it is compressed to this previously determined thickness. Following this compression, there are various processes that may be used on the copper to obtain the desired consistency or hardness. For electrical applications it is desired that the copper be relatively soft to increase conductivity in the material. This process, as previously mentioned, is used for more critical applications because the rolled copper is a much more evenly distributed and consistent copper than ED copper. Figure 2.11(a) shows a cross section of ED copper, while Figure 2.11(b) is rolled copper. You can see how the rolled copper is in very even layers, but the ED copper does not exhibit this obvious uniformity. This is not to say that ED copper is not a good conductor at microwave frequencies, but when critical applications are involved, it may be wiser for you to use a rolled copper laminate.

Now that we have the weight and fabrication of the copper defined, the next logical step is to put the copper on a microwave laminate. All microwave laminates, whether using ED or rolled copper, are put together,

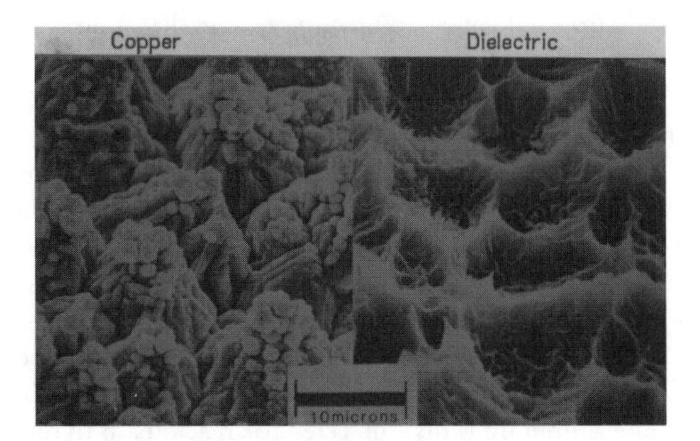

Figure 2.11(a) ED copper. Courtesy of the Rogers Corp.

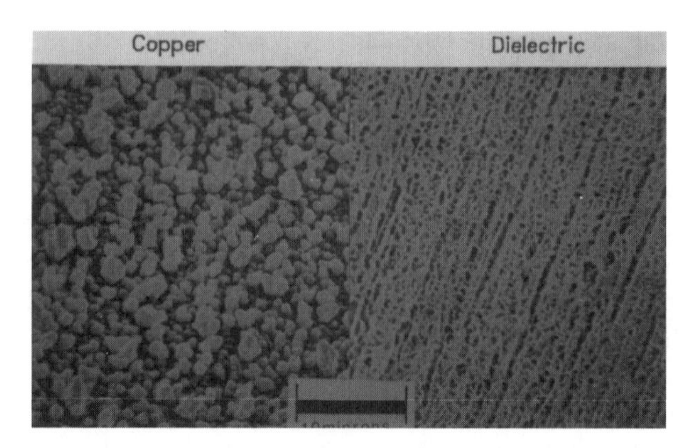

Figure 2.11(b) Rolled copper. Courtesy of the Rogers Corp.

as the name implies, by a lamination process. This basic process is familiar to all of us in one form or another. If you have ever had a driver's license or Social Security card or business card laminated, you know what we are talking about. Consider the steps involved in laminating one of your cards.

- The card is placed between two plastic laminating sheets.
- A press is heated to the lamination temperature.
- The card and laminating material are placed between the heated plates of the press and pressed together.
- The card is now laminated and would probably be partially or totally destroyed if you attempted to separate the laminations.

This is the same basic process that is used to put the copper on a microwave laminate. One very important task must be performed before the copper can be laminated to the dielectric material: a process of roughing up the surface of the copper so that it will "stick" to the dielectric. During the electrodeposition process, a certain amount of roughness is encountered. There usually needs to be additional roughing for good adhesion. Rolled copper normally is very smooth when completed and thus requires more roughing. This surface roughing is generally accomplished by means of an acid etch. This is similar to the roughing necessary when you iridite bare aluminum. The aluminum is first subjected to an acid bath to rough up the surface so that the iridite solution will "stick" to the surface of the aluminum. The same type process is used for copper prior to the lamination process.

When combining copper with a dielectric material, the same ingredients are used as with the business card—heat and pressure. With these two ingredients present and controlled, a highly reliable and efficient microwave laminate will result.

Now that we have covered cladding for laminates, let us discuss metallization for substrates. The formal definition of metallization is "the deposition of a thin film pattern of conductive material onto a substrate to provide interconnection of components or conductive contact for interconnection." A basic definition of the term is very simply the attachment of a metal to a substrate by means of depositing techniques. When referring to metallization on ceramic substrates (as used in microwave applications), we usually call the process by which we attach the metal to the substrate *sputtering.*

Sputtering is a highly controlled method of coating one material with another (for example, a ceramic substrate coated with a metal). The idea of sputtering is one of energy. The material that is ultimately deposited is literally blasted from a target material by high-energy gas ions. Normally, when you think about putting one material on another, you think of a process in which application is direct, as in painting or a lamination, as we discussed earlier regarding ED and rolled copper. This, however, is not the case with sputtering. The method of application is more of an *indirect* process. To understand this statement, and sputtering, consider the following example.

Suppose you had a cardboard box and placed balloons in one end of it as shown in Figure 2.12. The balloons are put in the box side by side so that there is a tight fit. The balloons are considered to be the target. If we now place a hose so that it is in direct line with the center of the balloons and send out a short burst of water, we will simulate the action of sputtering. (The hose and short burst of water represent the gas ion that causes the target material to be placed on the substrate.) The burst of water strikes a balloon (atom) and tries to push it to the back of the box. This cannot be accomplished because of the box itself and the other balloons holding it in place. Instead, the balloon compresses under the pressure of the water burst. When it comes back to its original spherical shape, it transmits the impact force to the other balloons around it. The force works its way to the front row where the balloons are no longer restricted, but are free to move. They move out in the direction from which the water burst came.

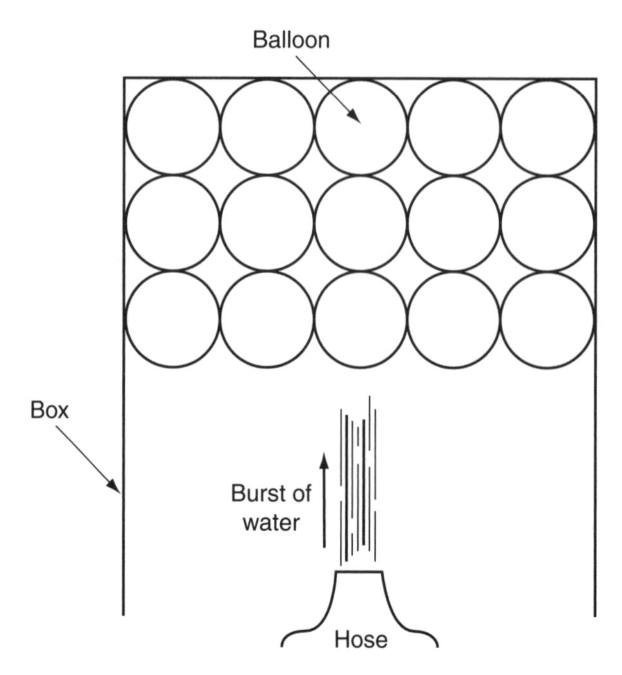

Figure 2.12 Sputtering example.

With the example presented in mind, you can look at Figure 2.13 and see how a basic sputtering process works. The target consists of the material to be deposited on the substrate (gold, for example). Notice that it has a negative (−) charge. The gas ion (usually argon) that bombards the target is shown in the center. The ion hits the target and breaks material loose just as we illustrated with the balloons. This material (atoms) travels through the vacuum to the positively (+) charged substrate. The layer deposited can be very carefully controlled by adjusting the voltage applied to the plasma (ion) and the time of exposure of the substrates.

There are two terms that may arise if you become involved with sputtering or plan to investigate further into its operation: *plasma glow region* and *dark space region.* Figure 2.14 shows where these regions are in the sputtering setup. The plasma glow region is produced because the target is negatively charged, which results in electrons being injected into the gas (usually argon) that is around the target. Because of the voltage applied to the system, the electrons are accelerated toward the positive charge on the substrates and substrate holder. As the electrons travel toward the positive side, they may collide with a gas molecule. When this happens, they give up part of their energy and leave behind an ion and an extra free electron. The cu-

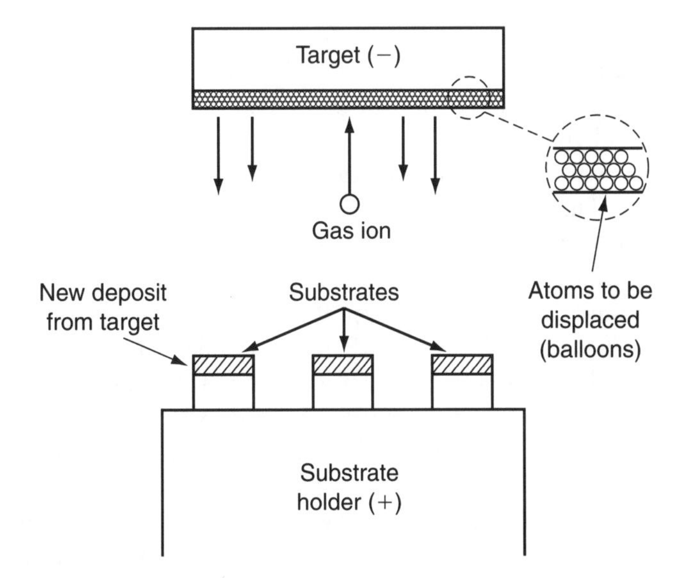

Figure 2.13 Sputtering process.

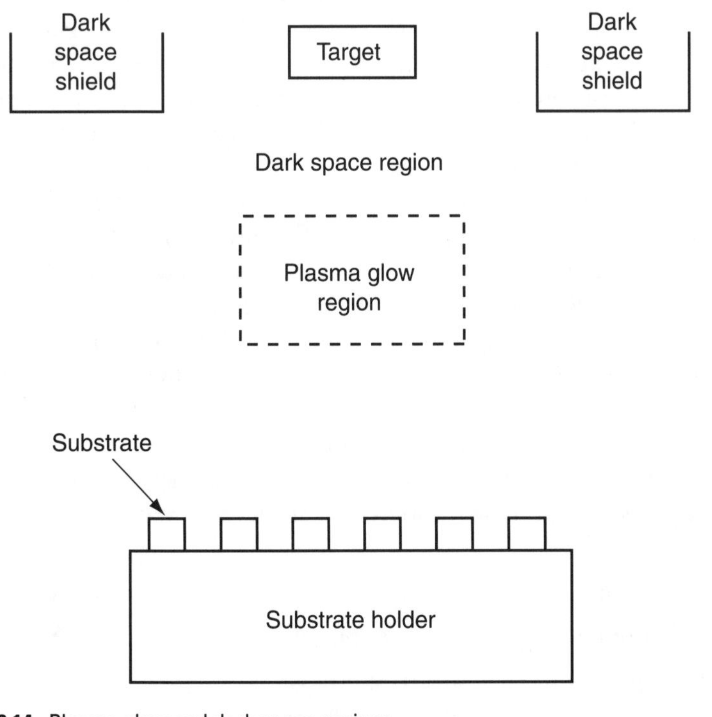

Figure 2.14 Plasma glow and dark space regions.

mulative effect of this phenomenon is a self-sustaining glow discharge. This ionized area actually heats up and glows. This effect can be viewed in a fluorescent tube. The only difference is the gas used (argon for sputtering, neon for fluorescent lights).

This glow will not occur close to the target, and this is called the dark space region. This is because the further the electrons travel, the more chance they have to strike and ionize the argon (gas) molecules. Therefore, very little ionization takes place close to the target; there are no collisions, no plasma glow, and thus a dark area appears.

The *dark space shield* shown in Figure 2.14 takes advantage of the dark space region. This shield (or ring) is placed around the target to ensure that a minimum amount of target material is wasted by being emitted from the side or back of the target. With the shield in place, the maximum amount of target material is sent to the substrate, where it is intended to be.

Although sputtering has recently become more prominent in microwave applications, it is not a new process by any means. The idea of sputtering was first presented in 1852 by Sir William Robert Grove. At that time, he called the process *cathode disintegration.* In 1909, F. Stark gave the first true and accurate description of the process, and in 1921 Sir Joseph John Thompson christened the process *spluttering.* The term *"sputtering"* resulted when a published paper in 1923 dropped the *l.* In 1925, sputtered films were investigated for use by Western Electric for commercial application. In 1928, Western Electric used the process to manufacture phonograph records and contacts on microphone transmitters. Until these applications, sputtering was used mostly for decorative effects. These early processes were dc sputtering and were useful only for metallic films. This, of course, is the process used for depositing metal on microwave substrates. In the early 1970s, RF sputtering was developed and allowed nonmetals to be sputtered. Advances in vacuum technology have further improved the quality and reproductivity of sputtered materials. Today, the major uses of sputtering are in the electronics industry.

Some of the combinations sputtered for microwave use are shown below. Note that in some cases, we say that the materials can either be sputtered or plated. All of these substrates are on alumina material.

Material A:

- Chromium (Cr)—adhesive-sputtered;
- Copper (Cu)—conductor-sputtered;
- Gold (Au)—passivation layer plated.

Material B:

- Chromium (Cr)—adhesive-sputtered;
- Gold (Au)—conductor-sputtered or plated.

Material C:

- Titanium tungsten (TiW)—adhesive-sputtered;
- Gold (Au)—conductor-sputtered or plated.

You will notice that none of the materials (substrates) listed is gold applied directly on the alumina substrate. This is because gold will not stick (or adhere) to the ceramic material. To put gold directly on the ceramic material would be similar to using superglue on wood. The glue soaks into the fibers of the wood, and there is no adhesion. Similarly, there is a migration of the gold into the substrate and no clear boundary between the substrate and the gold. This is a common fault with power transistors where gold metallization is used without the proper buffering and adhesive layer. This migration does not form a bond and a simple piece of a tape will remove the gold from the substrate. For this reason, an adhesive layer is needed (chromium or titanium tungsten in the examples shown).

Examples of usage for each of the materials listed are:

- *Material A:* For applications with solder;
- *Material B:* For applications where epoxy or bonding techniques are used;
- *Material C:* For high-temperature applications.

2.2.6 Coefficient of Thermal Expansion (CTE)

This parameter gets overlooked many times when a designer chooses a material for a particular application. The coefficient of thermal expansion (CTE) is the amount that the material changes mechanically with temperature. This parameter, expressed as parts per million per degree Centigrade (ppm/°C), is as significant a number for ceramic-filled materials as it is for the PTFE/glass materials. To understand just how important this parameter is, consider that the CTE for aluminum is 28 ppm/°C (this metal is used because many of the cases for microwave circuits are fabricated from aluminum). If we can have a material that has basically the same CTE as the case in which it is placed, we stand a much better chance of the circuit operating properly and remaining intact over a wide range of temperatures.

We stated that the CTE for ceramic-filled material and PTFE/glass materials was significant. This is true but for entirely different reasons. The CTE for PTFE/glass materials is 130 ppm/°C, while those for the ceramic-filled material is in the area of 24–25 ppm/°C. Thus, it can be seen that the ceramic-filled materials more closely match the CTE of the aluminum as opposed to the PTFE/glass materials. This parameter comes into play only when the microwave circuit you are designing is to operate over a wide temperature range. If the circuit is to only be in a basic ambient temperature environment, there is no need to concentrate on the CTE of the material, as it will probably not be a problem for your circuit.

2.2.7 Peel Strength

The construction of a microwave laminate will help to understand the peel strength parameter. The actual laminates consist of the material itself with copper on both sides. There are also cases where there may only be copper on one side of the material. Regardless of the number of sides that have copper attached to them, the peel strength comes into play in all cases.

Peel strength is the amount of force required to separate the copper from the dielectric material. It is expressed in pounds per inch (lbs/in) and is an important parameter that should be as high as possible to ensure that the copper stays attached to the dielectric material. The copper must stay attached during the manufacturing process, after it is sold, during fabrication, and when the material is in a circuit or system being used. The last thing that designers need is to have their beautiful circuit literally fall apart when it gets out in the field. Typical numbers for peel strengths are 8, 10, 12, and 15 lbs/in.

An important area where peel strength comes into play is during a soldering operation for the circuit board. When a soldering iron is applied to a piece of copper attached to a dielectric material, that copper must remain attached as securely as it was before the iron was applied. If the peel strength of the material is too low, the copper will dislodge from the material and the circuit will not operate properly. A good test for peel strength is to etch narrow transmission lines (approximately 0.015″ wide) on a piece of material and apply solder up and down the entire length of the transmission line. The idea is to heat up the entire line to see if the copper will remain attached to the material. If the circuit board material can withstand this test, the peel strength is more than adequate.

2.3 Microwave Material History

Before we deal with specific types of laminates and substrates, let us review their history. The history of laminates and substrates for eventual microwave usage began in the 1950s. Early experimentation used low dielectric constant plastics in a microstrip configuration, but these first attempts were not very successful because the effects of radiation and intercircuit coupling were not taken into consideration.

Polystyrene was found to be a uniform low-loss material but was shown to crack or soften at temperatures above ambient. Attempts to harden and reinforce the material by adding glass fibers destroyed many of the desirable properties that it originally had. As an alternative to polystyrene, PTFE/glass cloth laminates were considered. This material was originally developed for other uses and had some rather unpredictable dielectric properties, but it fulfilled many of the necessary requirements for a microwave laminate.

By the early 1960s, PTFE/glass cloth laminates had their dielectric constant reduced from their original 2.65–2.75 to the more popular and usable 2.55 of polystyrene. With this advance, the PTFE/glass laminate replaced the styrene product and eliminated many problems. By reducing the glass content even further, the dielectric constant was reduced to 2.45 and the first industry-standard laminates were created.

Dielectric constants were further reduced by the introduction of irradiated polyolefin (2.32) in the 1960s. This is a uniform and reproducible material that does not crack like the polystyrene. It does, however, have some rather serious mechanical instabilities, especially when temperature cycles.

PTFE/glass microfiber laminates were also introduced in the 1960s. The original material duplicated the dielectric constant of polyolefin (2.32) but had far superior dimensional stability and a much greater temperature range. A dielectric constant of 2.2 was achieved in the microfiber material by, once again, reducing the glass content.

A very popular dielectric material in the late 1960s was PPO (polyphenylene oxide). This material was touted as the ultimate in microwave laminates when it was first introduced. It had the ideal dielectric constant of 2.55, had very uniform and reproducible electrical properties, and was very stable both dimensionally and over temperature. Why is PPO not the only material used in microwaves today? PPO had a very large problem chemically: not only would etching solutions make the edges of boards very sticky (something like a piece of pizza coming out of a pan if touched to some other surface), but it had a habit of developing fine cracks at unpredictable times after the material was attached to a case. These cracks were attributed to improper processing and usually occurred in the drilling process. The manufacturers of PPO specified that the material should not be punched but drilled. Most people adhered to this recommendation but still encountered problems. The reason was that in order to drill PPO, you needed either a new drill bit or one that had just been sharpened. Anything less than that would be considered dull, and the properties of the material would be such that the bit acted as a punch and the fine cracks would show up at a later date. Usually, these cracks occurred after the circuit had been tested and was well on its way to being shipped. Needless to say, PPO finally disappeared from the microwave laminate scene.

During the mid- and late 1960s, the high-purity (99.5%) alumina substrate material arrived on the microwave scene, and microstrip technology began its rapid rise to prominence in the microwave field. Its high-dielectric constant (9.9–10) and smooth surface finish resulting in low-loss circuits made it an ideal material for high-density microstrip circuitry. Attempts were made in the 1960s to fabricate high-dielectric laminate that would compete with alumina substrates, but most were inconsistent and unreliable or had fabrication difficulties. A glass-cloth-loaded PTFE material with a dielectric constant of 6 was the only success in a long line of attempts. This material is still available today.

During the 1970s, microwave laminates and substrates really came of age. This was due partly to the new technologies that made more consistent and reliable materials available and partly due to the new systems being designed in industry that put pressure on laminate manufacturers to produce these materials.

Alumina substrates increased purity (99.6 and 99.7) during the 1970s and had finer particle size. Techniques for sputtering and deposition are refined so that thinner, more consistent metallization is possible. At the same time, laminates were moving in both directions regarding dielectric constant. They were going lower (2.17) for applications in the millimeter-wave range and higher (10.2) to give the alumina market competition. The road to these new laminates was not smooth, however. The first high-dielectric material that was released worked rather well on the bench for breadboard or unstressed applications. (By *unstressed*, we mean no extreme temperatures or humidity.) When subjected to these atmospheric factors, there were great inconsistencies in the material. Also, many times the copper lines would lift off the material if a soldering iron was touched to them. All of these conditions, of course, were unsatisfactory and made the microwave industry wary of anyone who came to them with a high-dielectric microwave laminate. This persisted for a number of years, but, as in all things, time (and extensive research) heals all wounds. By the latter portion of the 1970s, the material was reintroduced, and three separate manufacturers were then supplying it. This time it proved to be an excellent material and found many areas where it replaced alumina as well as many new areas of application.

The 1980s were a time when the microwave laminate made its strongest advances: materials that had better temperature properties, pure PTFE laminates with electroplated copper attached, materials that were tailored to more popular dielectric constants (2.94, for example), the addition of metal attached to the laminates (aluminum, copper, or brass), the lamination of two dielectric materials with a metal plate between them, and the addition of a resistive layer under the copper sheet to aid in making distributed element resistors. All of these advances have greatly enhanced the microwave laminate industry and made it a very sophisticated technology.

The 1990s brought about even more advances in the microwave material industry. During that time, the military markets were dwindling and the commercial markets were rising faster than many people could keep up with them. The idea of only a Teflon-based material being used for microwave circuits was vanishing rapidly. Also, cost suddenly became a primary issue, something that had not been that big of an issue up to this point. So, the 1990s were truly an exciting time for both the microwave material manufacturer and the microwave designer.

This dramatic change in philosophy brought about many new types of microwave materials in the 1990s. Suddenly, non-PTFE materials, thermoset polymer composites, and even polyester materials appeared on the scene and all were doing a credible job of providing low-cost, reliable microwave circuits. The 1990s proved to be the decade where the most remarkable advances were made in microwave materials.

As a result of the history of laminates and substrates, the microwave industry has a broad cross section of materials that not only can be used for microwave circuits but also can have the base plate of the final case ordered as an integral part of the material. The designer should be familiar with the many different types of materials available, analyze their properties, compare advantages and disadvantages of each, and choose the proper material for the application at hand. The days of a "pet laminate" should be over. Probably the most discouraging and senseless statement for a person versed in laminates and substrates to hear is "we have used this material for years, so we might as well use it again." With the choices available today, there is no need to be strapped to a material just because you used it before or because you have a large supply of it left over from the last job. A brief list of some of the materials available is given here.

Material	Dielectric Constant
Woven PTFE/glass	2.94
Woven PTFE/glass	2.55
Woven PTFE/glass	2.45
Woven PTFE/glass	2.33
Woven PTFE/glass	2.17
Nonwoven PTFE/glass	2.33
Nonwoven PTFE/glass	2.20
Ceramic-filled PTFE/glass	6.0
Ceramic-filled PTFE/glass	10.2
Alumina	9.0 to 10
Pure Teflon	2.1

2.4 Teflon Fiberglass Materials

A standard material for microwave applications for many years was Teflon-based. It was, and still is in some areas, an excellent choice for high-frequency circuit board materials because of its uniform construction. Recall that the high-frequency energy propagates through the material in order to perform certain functions. If there are obstructions in the material, the energy slows down and is not as efficient. Teflon material allows an unobstructed trip through the material. As good as Teflon material is for the electrical portion of the microwave applications, it is almost as bad for the mechanical portion of those same applications. Teflon is a very soft material that will "cold flow" if pressure is applied to it. That means the material changes its thickness dimension under pressure and does not always come back to its original dimension when the pressure is removed. That is a problem because the dielectric thickness, as we have stated, must remain intact and be held to very close tolerances to have the circuits perform properly. So, mechanically, pure Teflon is not one of the more acceptable materials to use. The key word here is *pure.* Most of the Teflon applications in microwaves are not pure Teflon. They are Teflon-based materials with fillers to make the material more acceptable mechanically and also change the characteristics electrically. The fillers usually are fiberglass, although recent advances use other materials.

The first Teflon-based material we will look at is the Teflon fiberglass material that was the first material to be used for RF and microwave applications. This material is usually referred to as PTFE/glass material, which stands for polytetrafluorethylene and has a fiberglass added to it. The fiberglass is used to reinforce the Teflon and make it more structurally rigid. It does this very nicely, but it also changes the characteristics of the original Teflon material. Consider the fact that we had a pure material (Teflon) and added another material to it. What we have done is to add impurities to a material. This, naturally, will change some of the parameters of that original material. When we add fiberglass to pure Teflon, we increase the dielectric constant of the material but we also increase the losses in that material. The dielectric constant of pure Teflon is 2.10 (you will see 2.08 in some reference books). When you add fiberglass to this, you can raise the dielectric constant to approximately 2.60 before the losses become too large and the addition of more fiberglass will only make the performance worse.

There are two methods used to place the fiberglass in a Teflon material: woven and nonwoven. The first to be discussed is the woven structure. This type of structure looks like the fabric of your clothes if they were magnified. Figure 2.15 shows a basic structure for such a material. It can be seen from the figure that there is a base of Teflon with the fiberglass strands woven throughout the material to form a nice structure that performs well in RF and microwave applications.

We stated that the parameters are changed when we add fiberglass to the pure Teflon base. To illustrate this, consider Table 2.3, which shows three woven PTFE/glass materials with three very important parameters compared: dielectric constant, dissipation factor, and peel strength.

It can be seen from Table 2.3 that the parameters will change as more fiberglass is added. The materials shown indicate that Material 1 has the lowest amount of fiberglass, with Material 3 having the most fiberglass added to it. With this in mind, you can see that the dielectric constant has indeed increased with the amount of fiberglass but the loss has also gone up. Another point to make here is that the peel strength has increased when you

Figure 2.15 Woven glass laminate construction.

add fiberglass to the pure Teflon. Remember that Teflon is the material that is also used to keep your eggs from sticking to the frying pan. So it should not be too hard to see that the less effect the Teflon has on the material, the better the copper will adhere to the material. This can be seen in the table as an increase in the peel strength when the dielectric constant reaches 2.50. Therefore, it can be seen that the amount of fiberglass has a significant effect on many parameters for the RF and microwave materials.

Table 2.3

Characteristic	Material 1	Material 2	Material 3
Dielectric constant	2.17 ±0.02	2.33 ±0.04	2.50 ±0.04
Dissipation factor	0.0009	0.0015	0.0018
Peel strength	8 lb/in	8 lb/in	12 lb/in

Table 2.4

Characteristic	Material 1	Material 2
Dielectric constant	2.20 ±0.02	2.33 ±0.02
Dissipation factor	0.0009	0.0012
Peel strength	15 lb/in	15 lb/in

The second type of PTFE/glass is the nonwoven material (also called *microfiber*). This is a Teflon-based material in which the fiberglass particles are added in a random fashion and not the woven structure we saw in Figure 2.15. This type of structure produces a material with a dielectric constant variation of less than ±1% and low dissipation factors. It is also very stable, dimensionally being able to withstand temperatures up to 550°F (288°C) without warping.

If we look at laminates (materials) with various amounts of fiberglass in a nonwoven structure and use the same parameters as we did in Table 2.3, we will have Table 2.4 for nonwoven materials.

It can be seen from Table 2.4 that the same pattern is realized as in Table 2.3 with the woven material. As the fiberglass content increases, the dielectric constant and dissipation factor both increase. The main difference shown in Table 2.4 is the fact that the peel strength is good for both materials and does not increase or decrease with the fiberglass content.

We stated previously that the introduction of fiberglass into a Teflon base will increase the dielectric constant of the resultant material. We also stated that as we increase the fiberglass content, the dissipation factor also increases to a point where the losses are very high above a dielectric constant of about 2.60.

To allow the continued use of Teflon for higher dielectric constants, the filler must be changed. The material that is used as a filler for higher dielectric constants with reasonable dissipation factors is one that finds many applications throughout the field of RF and microwaves. This material is *ceramic.* With the ceramic-filled PTFE, it is possible to obtain dielectric constants that range from 3.0 to 6.0 to 10.2. These materials find many applications for both high and low frequencies. To show how the ceramic-filled PTFE materials compare to one another and to the PTFE/glass materials previously discussed, refer to Table 2.5, which shows the three dielectric constants that are available as standard materials for many commercial RF and microwave applications.

Table 2.5

Characteristic	Material 1	Material 2	Material 3
Dielectric constant	3.0 ±4	6.0 ±5	10.2 ±0
Dissipation factor	0.0013	0.0025	0.0035
Peel strength	6–8 lb/in	6–8 lb/in	6–8 lb/in
Coefficient of thermal expansion	(20–25 ppm/°or all three)		

Notice the coefficient of thermal expansion numbers in Table 2.5 and how close they are to the CTE for aluminum.

2.5 Non-Teflon Materials

For many years, the only materials that were available for RF and microwave applications were those with a Teflon base, the PTFE laminates. These materials worked well for a long time and did a variety of tasks. With the change in markets in recent years, there has been the birth of a new line of materials that do not use Teflon but are reinforced hydrocarbon/ceramic laminates. Hydrocarbons are defined as any compound containing only hydrogen and carbon. Materials such as benzene and methane are common hydrocarbons. This is a different combination that has found many applications, especially since the common ceramic material is also used. We previously discussed the advantages and uses of ceramic materials in RF and microwave areas.

There is a very important property of these hydrocarbon/ceramic materials that make them more appealing than PTFE-based materials for some applications. There is a property of PTFE that can cause some problems if used for certain designs. This property is a *transition region* at 19°C (68.2°F). The dielectric constant is rather well behaved on both sides of the transition region, but the drop in dielectric constant is where the problem occurs. This is a natural property of Teflon and is something that has been tolerated over the years since there were no materials developed to eliminate the transition.

The curve for the hydrocarbon/ceramic material is a much more consistent curve that allows the material to have a predictable dielectric constant over a wide range of temperatures. There is no sharp drop in the dielectric constant, so a designer can use some straightforward relationships to design

Table 2.6

Characteristic	Material 1	Material 2
Dielectric constant	3.38 ±0.05	3.48 ±0.05
Dissipation factor	0.002	0.004
Peel strength	6 lb/in	5 lb/in

a circuit that will perform as expected. One area where this type of material is very valuable is in the design of filters. There are many components in microwaves, such as filters, whose construction is a series of resonators and quarter wave transmission lines. Their length depends very heavily on the dielectric constant of the material that is used. If this dielectric constant is varied over a temperature range, the filter response will also vary and could very well be completely out of the range to which the filter was designed.

Some of the properties of two hydrocarbon/ceramic materials are shown in Table 2.6. We show dielectric constant, dissipation factor, and peel strength for these materials.

These materials also have a coefficient of thermal expansion 46 to 50 ppm, which is relatively close to that of copper. When you think of this, you can understand how important it is to have the expansions of the copper and the material the copper is attached to be basically the same. This makes for a very stable material over a wide temperature range.

2.6 Thermoset Polymer Composites

Another material that has come about from increased commercial applications for microwave circuits is the ceramic thermoset polymer composite. Once again, notice the use of ceramics for a high frequency material. We said this term would come up over and over again, and it certainly has and will continue to for many areas of microwaves. These materials allow the material manufacturer to obtain a higher dielectric constant, good loss characteristics, and excellent material stability.

If you look at the type of material we are discussing and define all of the terms, you can get a better idea of what this material is. For example, a thermoset is defined as a material that becomes permanently hard and unmoldable when subjected to heat (certain plastics are thermosets); a polymer is a material that consists of many parts, either natural or synthetic; and

Table 2.7

Characteristic	Material 1	Material 2	Material 3
Dielectric constant	3.27 ±0.016	6.00 ±0.080	9.80 ±0.24
Dissipation factor	0.0016	0.0018	0.0015
Peel strength	(3 lb/in for all materials)		

a composite is defined as a material made up of many distinct parts. So what we have is a group of thermally combined materials with a ceramic base that will not soften when heated. This is an important property to have when you are operating in temperature extremes. The actual materials used in this type of laminate are, of course, proprietary, but they are obviously put together properly since they work so well. Table 2.7 shows some of the properties for these thermoset materials.

2.7 Polyester Materials

Probably one of the most recent materials to come on the scene, as of this publication, is one that seems unlikely to be associated with microwave circuits—polyester. That's right, the material that your clothes are made of. Once again, there is a proprietary process that makes this material compatible for use in RF and microwave circuits, but it does make a very nice high-frequency material that is finding its place in the industry for many applications. Table 2.8 shows the critical parameters of dielectric constant and dissipation factor for polyester material.

It can be seen that the polyester material has the parameters to be a viable consideration for many RF and microwave applications. Many areas of filters, antennas, couplers, and amplifiers have already found the polyester material to their liking.

Table 2.8

Characteristic	Material
Dielectric constant	3.20 ±0.04
Dissipation factor	0.0025

2.8 Alumina Substrates

We have referred to alumina substrates on various occasions throughout this chapter. Now it is time to explain them and where they fit into the microwave industry.

Until the 1970s, when high-dielectric PTFE material was introduced, alumina substrates were the only reasonably priced materials on which microstrip circuits could be built. These substrates were the familiar alumina, aluminum oxide, or Al_2O_3. Their popularity began around the mid-1960s when their purity went from the previous low of 96% to a much more acceptable 99.5%. Since then, it has gone to 99.7% and in some cases to 99.9%. The most widely used alumina substrates today use a purity of 99.6% or 99.7%. This increase in purity resulted in the "as-fired" surface finish dropping from a previous 8–10 microinch thickness to approximately 4 microinches. The term *as-fired* means that this is the substrate after its final firing without any additional machining or polishing. You might call these "rough" substrates.

The two items that are of importance in an alumina substrate and make it useful for microwave application are dielectric constant and dissipation factor. The dielectric constant of the substrate is affected by two factors: the preferred orientation of the alumina crystallites and the density of the material. Fine-grained alumina ceramic has a more random crystalline orientation than one with coarse-grained properties. This random orientation is difficult to regulate and depends very highly on an accurate process control during manufacturing. This type structure, however, is still preferred for the higher dielectric constant and lower dissipation factor. This is because the fine-grain structure results in a higher density in the material. The adhesion of circuit-forming films depends on the density of the grain boundary areas, and the fine-grain sizes produce a higher density that reduces *microwave scatter* caused by a coarse grain structure. This scatter is similar to a dissipation and increases the loss. Therefore, to have alumina substrates with a fine-grain structure is preferred for the optimum dielectric constant and dissipation factor. This guarantees consistent substrate properties.

Other than the dielectric constant and dissipation factor, the flatness of a substrate is another important factor when choosing the right material for your application. *Flatness* is the term used to define the overall *camber* of a substrate. (Camber is likened to the warpage of a material; see Figure 2.16.) This is usually expressed in the allowable inch out of flat per inch of substrate (0.001″/″). The term *camber* is important since flat substrates (or as nearly flat as possible) are necessary to ensure proper operation of vacuum

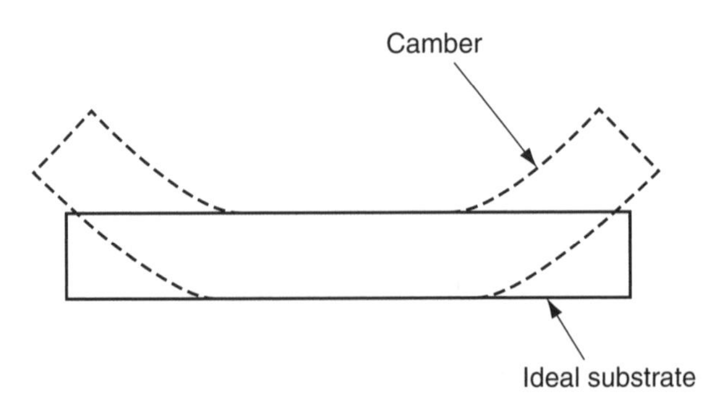

Figure 2.16 Substrate camber.

fixture during exposure, a good fit in holders during processing, and a good mask contact during the printing on the metallization.

Besides flatness, other terms used to describe the final finish of a substrate are *as-fired* and *polished.* We have previously described an as-fired substrate as the condition of a substrate after its final firing without any additional machining or polishing. You will recall that we called these "rough" substrates. These substrates, as might be expected, can have definite limits regarding flatness and dimension control. Most of the time, these limits make the as-fired substrates unsuitable for microwave application. When this occurs, you must specify *polished* substrates. These substrates offer excellent performance but usually are much higher in cost. These substrates are a fired material that is buffed smooth by mechanical grinding or lapping.

With the primary terms presented, we will close our discussions on alumina with a typical data sheet:

Material	Al_2O_3 (99.6%)
Dielectric Constant	9.90 ± 2%
Dissipation Factor	0.0001
Metallization	chrome-gold
Surface Finish	3 microinches (as fired)
Camber	0.001 ″/″
Thickness	0.025″ ± 0.00005″
Outside Dimensions	0.05″ × 0.5″ ± 0.002″

All of the terms in the data sheet above have been covered. This example should show you that alumina is an excellent substrate (electrically) for microwave applications. The only drawbacks in the material are its brittleness (it is rather like working with a dinner plate) and its difficulty to machine. Other than these factors, the substrate is an excellent choice for microwave usage.

2.9 Sapphire, Quartz, and Beryllia Substrates

These substrates probably should be put under a category of exotic. None of them could be called an everyday material that would be the first choice of a microwave designer.

Sapphire (also referred to as α alumina) is a material with a dielectric constant of approximately 11.0. This is about as close as we can pin down this material because it is not isotropic and dielectric constants ranging from 9.3 to 11.7 have been measured on it. The difficulties in the fabrication of alumina (cutting, drilling, etc.) are also encountered with sapphire substrates. The primary use of such a material is when a high thermal conductivity is desired. This would probably be the only real application of such a material for microwaves.

Quartz (SiO_2) is another brittle material that has the same fabrication difficulties as alumina and sapphire. Its dielectric constant is 3.78, which is somewhat higher than woven PTFE/glass material. This may be useful for higher frequency applications where smaller wavelengths need a low dielectric constant to keep the dimensions reasonable. The dissipation factor, unlike sapphire or alumina, is a relatively unimpressive 0.0015. This, you will recall, is equivalent to regular PTFE/glass laminates. Quartz would be used only in very special and rather loosely specified applications.

The last of the materials in this section is beryllia (BeO). It has a dielectric constant of 6.9 and a dissipation factor of 0.0002. This dielectric constant is put in the category of high-dielectric PTFE laminates ($\varepsilon = 6.0$) and in the alumina category of dissipation factor. Beryllia substrates require polishing because they have a coarse-grain structure, resulting in an inherently rough surface and a poor resistance to chipping. These substrates are normally used only where an extremely high thermal conductivity is needed. This is because the material is extremely expensive and has a high toxic hazard rating. This would normally be one of the last materials considered for typical microwave applications.

2.10 Material Specifications

For many years, MIL-P-13949 was used for microwave materials as a specification for all microwave circuits. In the early 1980s, the Institute for Interconnecting and Packaging Electronic Circuits (IPC) began work on a document that was designed to be specifically for microwave materials. The previous specification (MIL-P-13949) was not actually a specification for microwave materials but was adjusted for use with microwave materials. As has been stated, the IPC-L-125 is a specification for microwave materials and has worked very well since its inception. This can be further realized by looking at the IPC-L-125 name, Specification for Plastic Substrates Clad or Unclad for High Speed/High Frequency Interconnections. This reinforces the statement that this is a microwave specification.

You can see that the IPC-L-125, now in revision A form, is a complete specification by looking at the major headings for the specification shown here:

1.0 General
1.2 Scope
1.3 Type Designation
1.4 Dimensions
2.4 Applicable Documents
2.6 IPC
2.7 Department of Defense
2.8 ASQC
2.9 ASTM
3.9 Requirements
3.11 Specification Sheets
3.12 Terms and Definitions
3.13 Registration
3.14 Constituent Material
3.15 General Requirements/Acceptability
3.16 Physical Requirements
3.17 Chemical Requirements
3.18 Electrical Requirements
3.19 Environmental Requirements
3.20 Marking
3.21 Workmanship
4.21 Quality Assurance Provisions
4.23 Quality Conformance Evaluations
4.24 Classification of Inspections
4.25 Materials Inspection
4.26 Registration
4.27 Certification Testing
4.28 Quality Conformance Inspection

4.29 Statistical Process Control (SPC)
4.30 Test Methods
5.30 Packaging
5.32 Preservation
5.33 Packing
5.34 Marking
5.35 General
6.35 Notes
6.37 Ordering Data
6.38 New Materials
6.39 Supersession Data

It can be seen that this is a very comprehensive specification that works well for microwave materials. To see how easy it is to use such a specification, consider the following example of a number that is used to describe a material by using the references in the specification:

IPC-L-125 01 C1 A 0310 B C1 DH

The explanation of this number is as follows:

IPC-L-125	Specification Number
01	Specification Sheet Number (sheet 01 shown in Figure 2.17)
C	Permittivity (2.55)
2	Permittivity Tolerance (1%)
A	Pits and Dents (Grade A, 29 Max, in any 12″ x 12″ area)
0310	Dielectric Thickness (0.031″)
B	Thickness Tolerance (0.003″)
C1 DH	Metal Cladding Type and Weight or Thickness (Electrodeposited ½ ounce copper)

So it can be seen that this number has precisely described a piece of PTFE/glass material that is 0.031″ ±0.003″ thick, with a dielectric constant of 2.55 ± 1%, with ED copper of ½ ounce in weight. A typical data sheet from the specification is shown in Figure 2.17, previously referred to in the material number explanation. Various data sheets for microwave materials available on the market today are shown in Appendix F. As of this text, the IPC specification is in revision A.

July 1992 IPC-L-125A

IPC-L-125A

IPC-L-125A Specification Sheet

Specification Sheet # : 125/02
Reinforcement : Woven E Glass
Resin System : PTFE
Permittivity Test : 2.40 – 2.60
Permittivity Test Frequency : 1010 Hz
(Formerly MiL-P-13949 type GX)

Nominal Permittivity:
A – 2.40 E – 2.60
B – 2.45 X – As specified in ordering data
C – 2.50
D – 2.55

	Class 1	Class 2	Class 3 & Class 4
1. Peel strength, Newtons/m [lb/in], ave. min.	<0.75 mm [0.030 in.]/ ≥0.75 mm [0.030 in.]	<0.75 mm [0.030 in.]/ ≥0.75 mm [0.030 in.]	<0.75 mm [0.030 in.]/ ≥0.75 mm [0.030 in.]
A. <17 micron Thermal Stress	875 [5.0] / 1050 [6.0]	875 [5.0] / 1050 [6.0]	875 [5.0] / 1050 [6.0]
[<1/2 oz./ft.^{2}] Elevated Temp. (150°C)	700 [4.0] / 875 [5.0]	700 [4.0] / 875 [5.0]	700 [4.0] / 875 [5.0]
B. 17 micron Thermal Stress	700 [4.0] / 700 [4.0]	700 [4.0] / 700 [4.0]	700 [4.0] / 700 [4.0]
[1/2 oz./ft.^{2}] Elevated Temp. (150°C)	350 [2.0] / 700 [4.0]	350 [2.0] / 700 [4.0]	350 [2.0] / 700 [4.0]
C. 34 micron Thermal Stress	875 [5.0] / 1400 [8.0]	875 [5.0] / 1400 [8.0]	875 [5.0] / 1400 [8.0]
[1 oz./ft.^{2}] Elevated Temp. (150°C)	350 [2.0] / 1400 [8.0]	350 [2.0] / 1400 [8.0]	350 [2.0] / 1400 [8.0]
D. 68 micron Thermal Stress	1050 [6.0] / 1750 [10.0]	1050 [6.0] / 1750 [10.0]	1050 [6.0] / 1750 [10.0]
[2 oz./ft.^{2}] or greater Elevated Temp. (150°C)	350 [2.0] / 1750 [10.0]	350 [2.0] / 1750 [10.0]	350 [2.0] / 1750 [10.0]
2. Volume Resistivity, mohm.–cm., min.			
Condition F – Humidity	NA	10^5	10^5
Condition E – 24/150	NA	10^5	10^5
3. Surface Resistivity, mohm., min.			
Condition F – Humidity	NA	10^2	10^4
Condition E – 24/150	NA	10^3	10^3
4. Moisture Absorption, max. percent	NA	0.20	0.20
5. Dielectric Breakdown, kv., ave. min. (≥0.75 mm [0.030 in.], D48/50 + DO.5/23)	NA	20	20
6. Loss Tangent, max. (C24/23/50) εr>2.50	0.003	0.0025	0.0025
εr≤2.50			0.0023
7. Flex. Strength, kPa [kpsi], ave. min.			
(≥0.75 mm [0.030 in.]) Length	NA	82.7 [12.0]	82.7 [12.0]
Cross Length	NA	68.9 [10.0]	68.9 [10.0]

NA = Testing not applicable for this class or condition

Figure 2.17 IPC-125 data sheet.

2.11 UL Requirements

Today, with the introduction of many new types of materials to be used for microwave circuits, there is a need for an additional specification to be incorporated. That specification is from Underwriters Laboratory (UL) and is UL-94, which defines flammability ratings for the material. With previous materials, it was not necessary to define this parameter on a data sheet because the materials were made up primarily of Teflon and fiberglass. Today, with all of the composites and polyester materials that are available, this parameter must be specified for circuit reliability and safety of the overall system.

To understand what needs to be done, let us look at the testing procedure of a material sample to see if it conforms to the UL specification. We will generalize the procedure to a certain extent. If you are actually running the test, you should have the specification in front of you and follow it to the letter.

Procedure:

- Clamp the material from the upper 6 mm (0.236″), with the longitudinal axis vertical, so that the lower end of the material is 300 ±10 mm (11.8″ ±0.393″) above the horizontal layer. The sample to be tested is 50 x 50 mm (approximately 2″ x 2″) and has a maximum thickness of 6 mm (0.236″).
- A methane gas supply is used and adjusted to produce a gas flow of 105 ml/min with a back pressure less than 10 mm (0.393″) of water. For more details, consult specification ASTM D 5207.
- The burner should be adjusted to produce a blue flame 20 ±1 mm (0.787″) high.
- The test flame should be calibrated in accordance with ASTM D 5207 at least once a month and when the gas supply is changed or when equipment is replaced.
- Apply the flame centrally to the middle point of the bottom edge of the material so that the top of the burner is 10 ±1 mm (0.394″) below the point of the lower end of the material and maintain it at that distance for 10 ±0.5 seconds. If the material drips molten or flaming material during the flame application, tilt the burner at an angle of up to 45° and move the burner so that the material does not drip into the burner. After the 10 ±0.5 seconds' exposure, withdraw the burner at a rate of 300 mm (11.8″) per second to a distance of at least 150 mm (5.9″) and begin to measure the afterflame time, t_1, in seconds. Record t_1.
- When the afterflaming stops, place the burner again under the material and maintain the burner at a distance of 10 ±1 mm (0.393″) from the remaining portion of the material for an additional 10 ±0.5 seconds (be sure to keep the burner free from any dropping material). Remove the burner as indicated above and measure the afterflame time, t_2, and the afterglow time, t_3. Record both values.

Table 2.9

Criteria Conditions	V-0	V-1	V-2
Afterflame time, t_1 or t_2	≤10s	≤30s	≤30s
Total afterflame time ($t_1 + t_2$) (5 specimens)	≤50s	≤250s	≤250s
Afterflame plus afterglow ($t_2 + t_3$)	≤30s	≤60s	≤60s
Afterflame or afterglow of any specimen up to the holding clamp	NO	NO	NO
Cotton indicator ignited by flaming particles or drops	NO	NO	YES

NOTE: If it is difficult to visually distinguish between flaming and glowing, which it is many times, a small piece of cotton held with tweezers, approximately a 50-mm square (approximately 2″ x 2″), should be brought in contact with the area in question. Ignition of the cotton indicates flaming.

To see how this testing applies and how you can interpret the individual data sheets for microwave materials, refer to Table 2.9. This sets up material classifications according to the parameters that have been measured in the UP-94 specification.

Thus, it can be seen that there is a definite flaming of some materials that needs to be detected and passed on to any circuit designer who has an environment where this could be a problem. Consulting with UL-94 and using it for your particular application will be a tremendous asset to your overall circuit and system.

2.12 Choice of Materials

The question often is asked, "How do you choose an RF and microwave material?" To answer this, one must know many factors about the application of that material. It is not a question that can be answered by always saying material A is best or material B is ideal. You have seen from our discussions that there are many factors that come into play when looking at these materials. There are also many different types of materials available, and no one material will do every task that comes up. So the choice is one that must be studied before the final decision can be made. This is much different from years ago, when only one material was available. Many times, this material was not good for an application, but it had to be adapted since it was the only one you could get. The designer today is faced with the opposite dilemma: There are many, many materials available and they have to sort through all of the claims and advertising to determine just which material will do the job for them. Actually, it is a nice position to be in because if you do your homework and really look at the materials and their specifications, you can come up with a material that will do an excellent job for you and make your life much easier now and in the future.

When choosing a material, two terms will probably come up first. These terms are *dielectric constant* and *frequency*. Actually, these terms are very closely related. If you remember earlier in this chapter, we presented a chart that related dielectric constant and frequency (Figure 2.1). There were areas that overlapped, but you generally could begin to choose a material from that chart. The dielectric constant is important and related to frequency because the lengths of the transmission lines will be determined by the dielectric constant of the material and their frequency of operation. What you should look for is a dielectric constant that will produce "reasonable" length transmission lines. This is sort of a general way of saying it, but you should realize that for most applications, a transmission line that is supposed to be a quarter wavelength should not be 25″ long or even 0.010″ long. These would be very unreasonable lengths. So use some common sense and choose a material that will give you some good lengths.

The next parameter that is looked at is the *dissipation factor*. This, as we have said, is the loss characteristic of the material. If the operation of your circuit is dependent on having very low losses within the circuit, be sure to choose low dissipation materials for that application. If you can compromise on this parameter, you probably can use a variety of materials for your circuit. Dissipation factor for filters and antennas should be one of your primary parameters to look at when choosing a material.

Such mechanical properties as *peel strength*, *coefficient of thermal expansion*, and *dielectric thickness* are all parameters that should be looked at. The peel strength ensures that the copper is going to stay in place when the circuits are in operation. The coefficient of thermal expansion should be looked at very intently if your application is to be subjected to temperature extremes. Examination of this parameter and of the interfaces of materials, such as copper, aluminum, and the material itself, will ensure that there will be no broken connections or removal of ground planes during these temperature extremes. The dielectric thickness has been discussed many times in this chapter. You should look at this because it will determine the widths of the transmission lines you will be using and let you know if they will be easy to bend around corners if necessary. Also, the thickness will let you know if you will be able to construct that directional coupler with a very small gap between the main line and the coupled line.

There are obviously other parameters on a material data sheet, but those just mentioned are the important ones that should be considered for the majority of applications for RF and microwaves. It cannot be stated too strongly that it is very important to really look at the materials that are available and choose the one that is going to do the job for you. So the answer to the question of how to choose an RF and microwave material is, "very carefully." Remember that the best material that is available today is the one that will do the job for you and your applications.

2.13 Chapter Summary

In this chapter, we covered an area that is usually overlooked, or even ignored, by many people. They still consider the microwave material to be a printed wire board (PWB) and do not understand how important this material is. By now, the reader should understand that these materials are very important and should treat them with respect.

We began the chapter by defining many of the terms that relate to microwave materials. We looked at dielectrics, dielectric constant, dissipation factor, dielectric thickness, peel strength, copper weight, and anisotropy. Each term was defined and it was shown how they influence the operation of a circuit that is placed on the material.

Next, we discussed material requirements and emphasized that the very close tolerances for the materials on such parameters as dielectric constant and dielectric thickness make them very special materials. The next section followed with five types of materials. These were Teflon fiberglass materials, ceramic-filled PTFE, non-PTFE, thermoset polymers, and polyester. Characteristics for each were pointed out and investigated. An abbreviated data sheet for each material was also presented. A presentation of the "soft" substrates were followed by a discussion of alumina, sapphire, quartz, and beryllia "hard" substrates and some of their properties.

Finally, the very critical choice of which material to use was covered. This is a process that, as previously mentioned, should take just about as much time as the design itself. It is that important. Some of the critical parameters were outlined as well as some steps to help choose the proper material.

So, we have presented the actual materials that make the microwave circuits tick. These materials have come a long way in the few years that they have been around, and they will go much further in the future. Stay tuned for an update.

3

Metals

3.1 Introduction

When you put together a book on microwave materials and how to fabricate circuits from those materials, it may seem a little strange to include a chapter on metals. However, if you think about some of the comments made in the first two chapters and consider how important a metallic presence is in microwave circuits, then this chapter makes sense.

When microwave designers begin the task of designing a microwave circuit, they must be more than people who know about electronic circuits and how they react at high frequencies. They must also be part mechanical engineer, chemist, metallurgist, and sometimes they must even work a little magic in order to have a fine working circuit. This is why this chapter is so important to microwave designers or microwave technicians who will be working with not only the circuit boards but also the metallization on them and the cases into which they will go. It is worthwhile to familiarize yourself with the properties of the metals in this chapter. You will see them many times throughout your microwave career.

Metals that find wide usage in microwaves are aluminum, copper, silver, gold, indium, tin, and lead. The last three are more prominent as combinations in solder than in a pure metallic form. These are not the only metals used, but they are the most prominent. You will recall from our previous discussions, for example, how chromium (chrome) and titanium tungsten were used on substrates as adhesive to keep gold attached to the ceramic substrate. These are not widely used metals but are important in specific applications. We will concentrate in this chapter on metals that have wide applications throughout the microwave industry.

The question that arises at this point is what makes a metal suitable and acceptable for microwave usage? Two properties are of prime importance when considering a metal for microwave usage: (1) good electrical conductivity and (2) machinability.

Good electrical conductivity is important because the metal used in microwave circuits must carry high-frequency currents with as little loss as possible. This metal may be used as a conductor for a circuit (which is part of the ground plane). Regardless of the application, low loss (good conductivity) is necessary. Values of conductivity for the metals previously listed are given here:

Material	Conductivity (mho/meter)
Silver	6.30×10^7
Copper	5.85×10^7
Gold	4.25×10^7
Aluminum	3.50×10^7
Indium	1.11×10^7
Tin	0.877×10^7
Lead	0.456×10^7

The metals are listed in order of conduction; that is, silver is the best conductor, and lead is the worst of this group. This is not to say that lead is a poor conductor, only that it is poor compared to the rest of the metals listed. This, however, is not a problem, as lead is not used for metallization on substrates or for cases that must be a part of a critical ground plane; it is used in solder with various other metals. Its conductivity by itself, therefore, is not a critical item.

Machinability of metals is important because very few microwave systems will fit into standard square or rectangular spaces. Most have a cutout, an overhang, or a curved portion needed to fit a rounded area. For this reason, the metals must be sheared, drilled, milled, or generally shaped to conform to the necessary space. In order to be cost effective and efficient, the metals must be easily machinable.

We previously mentioned seven metals that find wide usage in microwaves. These will be covered in the following order:

- Aluminum;
- Copper;
- Silver;
- Gold;
- Miscellaneous metals—
  - Indium;
  - Tin;
  - Lead.

Before discussing each individual metal, it might be interesting to look at a periodic table of the elements and see where each of these metals falls. Figure 3.1 shows a table with the seven metals we will cover listed in their appropriate places. You can see that aluminum is by itself, away from the other six. It is in the category classified as *very active metals.* The remaining six metals are in the category of *soft metals.* The metals silver (Ag) and gold (Au) are also listed in a separate group called the *noble metals.*

The area termed *ferrous metals* is one in which we do not have any metals listed that may be used in microwaves. Even though we have discussed chromium (Cr) as an adhesive and nickel (Ni) as a plating material, these generally do not have the machinability or conductance properties of the soft metals.

3.2 Aluminum

Aluminum is usually first on the list of metals used in microwave applications. The majority of cases for microwave components are fabricated from aluminum. You see gold-plated aluminum cases, iridited aluminum cases, anodized aluminum cases, and just plain painted aluminum cases. Many "caseless" components in microwaves have aluminum ground-plane plates that enable the component to operate properly. The list goes on and on, but the message still comes through as to the importance of aluminum in microwaves. With a metal that is this important, it might be interesting to look into a little bit of its history.

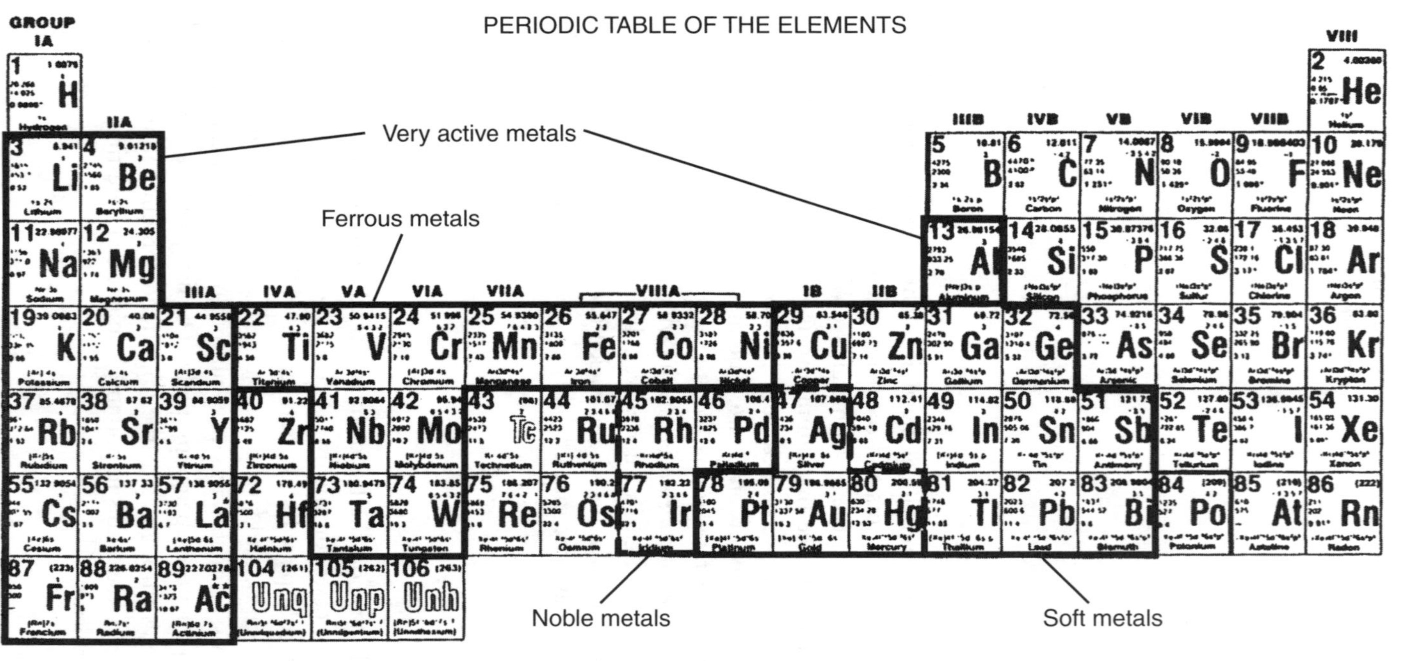

Figure 3.1 Metal locations. © Sargent-Welch Scientific Company, 1979.

In 1808, the existence of aluminum was predicted by the English scientist Sir Humphrey Davy. At that time, it was considered to be the most abundant metal on the earth's crust, but it was very difficult to extract from the earth. The first aluminum metal in the world was produced in 1825 from the Danish scientist H. C. Oersted. At that point, it sold for $160 a pound, something that may be rather difficult to grasp when you think of how inexpensive and widely used aluminum is today.

In 1854, Deville, a French chemist, helped the economic situation of aluminum somewhat by using sodium as a reducing agent for the metal. This brought the price down to a mere $100 a pound. This price, of course, meant that only a very few could afford anything made from this special metal. Emperor Napoleon III of France had his finest dinner spoons made of aluminum to impress his guests in the 1860s.

Charles M. Hall found a way to refine the metal inexpensively in 1886 that reduced the price to the range of $8 a pound. Hall, who was a student at Oberlin College, filed his patent on the process just ahead of Paul Herould of France, who had simultaneously and independently discovered the same process. Hall's process was bought in 1888 by the Pittsburgh Reduction Company, the first company to produce aluminum. The process which was purchased in 1888 is the same basic process used today in the current two-step process of aluminum refining.

Until World War II, the Aluminum Company of America (ALCOA), the successor to the Pittsburgh Reduction Company, was the only producer of aluminum in the United States. During the war, our aluminum production capacity was expanded greatly from the 1939 figure of about 164,000 tons. Although the majority of the plants to handle this production were built and operated by ALCOA, the latter's knowledge was made readily available to Reynolds Aluminum, the other prominent company.

The first cutback in aluminum production was made in December 1943, and by the end of 1944, the industry was operating at about one-half capacity. These cutbacks were needed simply because the supply of aluminum far exceeded the war needs. The huge industrial expansions that followed the war, however, more than made up for any cutbacks during the war. The price of aluminum has decreased significantly, even from the wartime prices. Following the war, in 1947, the price dipped to a record low of 15 cents per pound. The price, of course, did not hold at this level. Its present price of $2 to $3 per pound has been fairly constant for a number of years.

The only metal that has higher production quantities in the United States is steel. This is because of the wide application range and versatility of aluminum. Aluminum is a soft metal, although it is not listed as one in the periodic chart shown in Figure 3.1. You will recall that this group was called "very active metals." The metal is actually a soft metal that, almost contradicting logic, becomes so brittle at high temperatures that it can easily be powdered. The melting point of the metal is 660°C (1220°F), and it will boil at 2057°C (3734.6°F).

Aluminum is not the best conductor, but it has a very acceptable level of conductivity (3.50×10^7 mho/m as opposed to 6.3×10^7 mho/m for silver). Its softness enables it to be rolled into forms, cast, drawn, or stamped. It can, in its finished form, be milled, sheared, punched, or drilled. In short, it is a very machinable metal.

A general description of aluminum is given here:

Symbol	Al
Atomic weight (referenced to carbon 12)	26.98
Specific gravity	2.7 (a little more than 2½ times as heavy as water)
Color	Silvery-white with a bluish tinge
Properties	Soft, easily shaped, resists corrosion, nonmagnetic, good conductor of heat and electricity
Chief ore	Bauxite (named after the town of LesBaux in southern France where it was first mined)

This is aluminum: a soft and machinable metal with an average conductivity that finds applications in microwaves as cases and ground planes. The one property that accounts for its wide usage is its light weight.

3.3 Copper

If the first metal you think about for use in microwaves is aluminum, the second would surely be copper. This is not a new metal by any stretch of the imagination. It was probably the first metal ever used by man. Objects of hammered copper 8,000 years old are known from Egypt, and specimens of cast copper from Egypt and Babylonia date from 4000 B.C. The metal was given its name by the ancient Romans, who called it *aes cyprum* (metal of Cyprus), as the island of Cyprus was the chief source of copper for the Romans. Later they called it *cuprum.*

When the first Europeans came to the New World, they found the American Indians using copper for jewelry and decorations. The copper that these Indians used was from the Lake Superior area, deposited there by the action of the Ice Age. At that time, native copper was broken off from exposed lodes in the Great Lakes region and scattered southward over an area of over 70,000 square miles.

The mining of copper and the development of the industry began in the 17th century. In 1664, the first copper mine was established in the United States in Lynn, Massachusetts, and produced copper pins made from native copper in 1666. In 1780, Paul Revere established a foundry after discovering the secret process used by the English by which copper could be made malleable enough to be hammered when it was hot. The government loaned him $10,000 to buy a site for the foundry in Canton, Massachusetts.

In 1846, the first successful copper mine opened in the Lake Superior, Michigan, area. This is the time when the first gold mines were developed in Butte, Montana, which did not see copper mines until 1875. Mining was so extensive in this area that the entire city of Butte is undermined by the 700 miles of passages that run approximately 4,000 feet into the ground.

Copper is one of the metals in the category of coinage metals because it is used extensively in the manufacture of coins. There is a very high percentage of copper in coins today, much higher than in the past. Its price has risen sharply due to its scarcity, but this should not be a factor in influencing anyone to change his or her copper applications to another metal. The world is in no danger of running out of copper. Also, many applications do not require a pure 100% copper. Such alloys as brass or bronze do an acceptable job in many cases.

Copper fulfills our two initial requirements for a metal to be used in microwaves: (1) its conduction is excellent (5.88×10^7 mho/m), second only to silver (6.3×10^7 mho/m); and (2) it can be drilled, tapped, sheared, sawed, and milled. One thing to remember when machining copper, however, is that special lubricants may be necessary because copper is harder than metals such as aluminum. A quick check will save you from gathering a pile of broken taps and drills.

Copper is also one of the easiest metals to solder. It can also be heliarced, welded, or brazed. This ease of connection makes it ideal for low-loss microwave circuits.

Today, copper is not used as much as in the past because of its cost. If you looked in a waveguide catalog, you would find extensive use being made of alloys of aluminum, magnesium, brass, and silver. These all make excellent conductors, are lightweight, and are lower in cost than copper. The laminate market, however, still uses 100% copper because it has found nothing to match copper's excellent conductive properties for the price.

A general description of copper is given here:

Symbol	Cu
Atomic weight (compared to carbon 12)	63.54
Specific gravity	8.9 (nearly nine times as heavy as water)
Color	Reddish brown (pink when freshly made)
Properties	Soft and easily shaped, becoming hard when cold-worked; good conductor of electricity and heat; resists corrosion by atmosphere and sea water
Chief ore	Chalcopyrite (a compound of copper, iron, and sulfur)

Copper, although not as widely used today as before, still has many applications in microwaves. Its contributions to the microwave laminate and substrate market are immeasurable, as you can see from our discussion in Chapter 2. Its versatility and properties should make it an important part of the microwave industry for many years to come.

3.4 Silver

A second coinage metal is silver. This one is probably more recognizable than copper because most coins show the silver. Copper is usually visible only on the edges after extended usage wears the silver plating away.

Use of silver dates back to the earliest of times. Ornaments of silver have been found in the Near East that date back to about 3500 B.C. The book of *Genesis* mentions silver as part of Abraham's wealth. The first recorded mining of silver in Europe, however, was no earlier than 500 or 600 B.C.

Despite the demand for silver throughout the world, very little was produced during the Middle Ages because Europe's supply of silver was limited and its mining techniques were not highly developed. This shortage diminished when, in the 1500s, silver mines were discovered in Mexico, Peru, and Bolivia. Silver was not discovered in the United States until the 1700s.

During the Civil War, silver dollars and many other silver coins disappeared from circulation because the demands of industry for the metal caused it to be more valuable than the silver coins. After the war, in 1865, large silver mine discoveries in the Rocky Mountains decreased the price of silver.

Another silver shortage occurred in the early 1900s when large amounts were sent to India to help avoid a collapse of its currency system. A silver-purchase act passed by Congress in 1934 directed the Treasury to purchase silver, resulting in a large surplus. This surplus did not last, however, because ever-increasing uses were found for the metal. In July 1965, the U.S. government changed the metallic content of dimes, quarters, and half-dollars, and the result is the silver coin with the copper edges that you see today.

The use of silver for microwave applications is almost entirely in plating. Waveguide and tuned cavities are made of a lightweight and inexpensive material (aluminum, etc.) and plated with a thin coating of silver to increase conduction of microwave energy. The coating is very thin because the microwave energy travels only on the "skin" of the waveguide or cavity due to the phenomenon known as the "skin effect." It is much more economical and practical, therefore, to machine the cavity or fabricate the waveguide and plate it with the skin-depth thickness of silver. The price of silver is not quoted in this text because it changes from day to day.

A general description of silver is given here:

Symbol	Ag (from the Latin *argentium*)
Atomic weight (compared to carbon 12)	107.88
Specific gravity	10.49 (10½ times as heavy as water)
Color	White
Properties	Soft and easily shaped; the best conductor of heat and electricity; resists corrosion by the atmosphere

3.5 Gold

Most of the statements made in the previous section on silver could be repeated here for gold, starting by saying that its use dates back many thousands of years. The use of gold predates modern civilization, and many ornaments that have been found in Neolithic remains verify this statement.

A major use of gold in microwaves is for plating. Stripline and microstrip circuits are plated. Flanges and threads of microwave components are gold plated, as is the entire case or substrate carrier plate, to increase conduction and prevent corrosion of base metals used.

Gold is an excellent metal for use in microwaves but is being replaced—and in many areas has already been replaced—because of its price. For many years, the price of gold was a nearly fixed figure, and any industry that had applications for it could plan its price and be assured of its accuracy. In recent years, however, the price of gold has varied from day to day (and sometimes hour to hour) and has jumped in price to a point where it is impractical in many cases to use it. Many circuits are now using tin-lead plating to save money, and connectors now use a stainless steel finish. These variations in plating have kept microwave components within a reasonable price range.

There are times when a gold plating is really the best metal to use for a particular application. When this is the case, the designer usually makes the case in sections and gold-plates only the base plate, for example, which will act as the ground plane for the circuit. The remainder of the case is iridited, and the circuit usually operates very satisfactorily.

A general description of gold is given here:

Symbol	Au (from the Latin *aurium*)
Atomic weight (compared to carbon 12)	196.96
Specific gravity	19.32 (nearly 20 times the weight of water)
Color	Golden yellow when pure; impurities cause various shades of yellow
Properties	Very soft and easily shaped; extremely resistant to corrosion; excellent conductor of heat and electricity; reacts to very few chemicals

3.6 Miscellaneous Metals

Under this category of miscellaneous metals, we will cover three metals that are in the "soft metal" group and their primary use is in solders. The metals are indium, tin, and lead. Figure 3.2 shows the location of the three metals on the periodic table. You can see that they are all in a group and thus exhibit much the same properties. Each, however, has something that it does well. Two alloys that find wide usage in microwaves will also be presented.

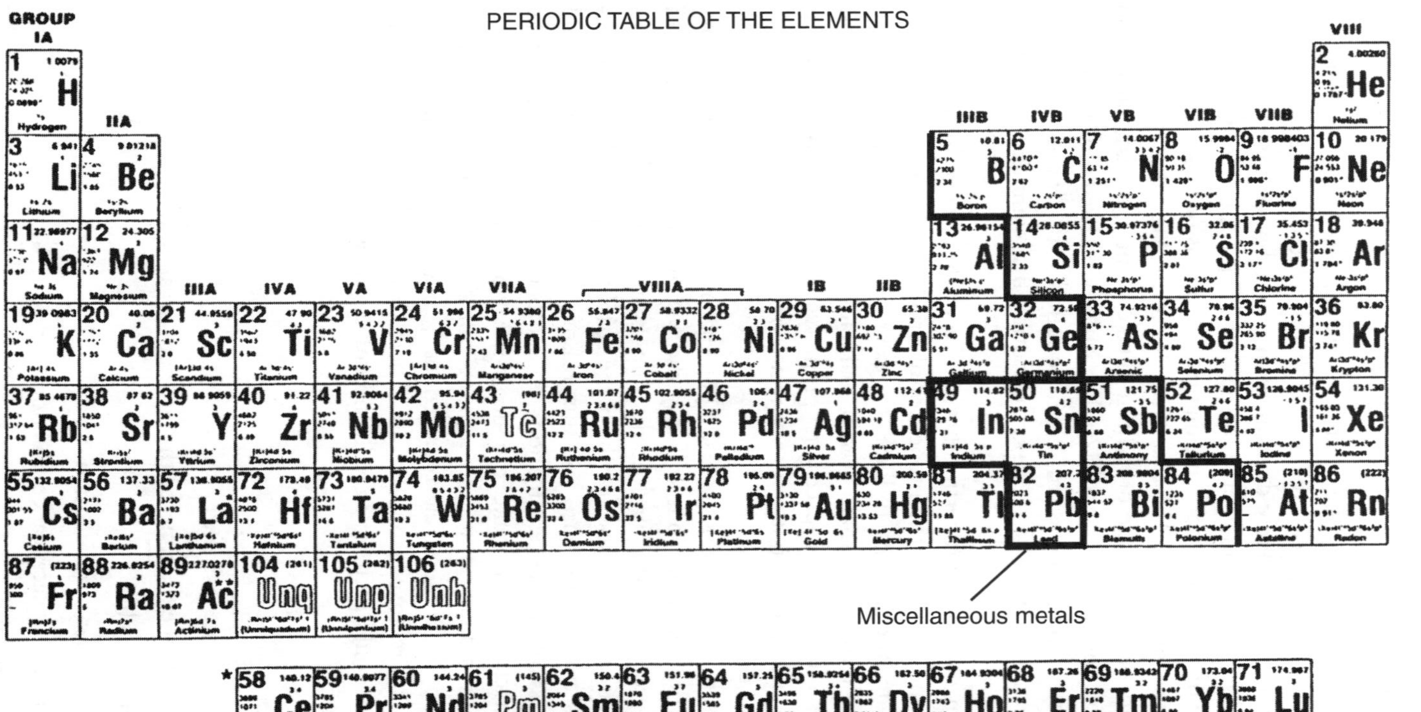

Figure 3.2 Soft metals. © Sargent-Welch Scientific Company, 1979.

3.6.1 Indium

When you consider some of the metals we have discussed thus far, indium is a relative infant. It was discovered in 1863 by two German scientists, Reich and Richter. Indium is widely distributed in nature but occurs in commercially practical concentrations only in the ores of zinc, iron, lead, and copper. Because of this, it did not have many applications but was a laboratory element. It was not until 1924 that Dr. William S. Murray decided to do something about the situation of indium to take it out of the laboratory and put it to work commercially. In 1926, Dr. Murray received a patent for his indium electroplating process. In 1934, he founded the Indium Corporation of America in Utica, New York, the world's first commercial producer of indium metal.

Indium has many properties that can be of great value in microwave application and in aiding the properties of other metals that could possibly be used in microwaves. Indium, for example, is so soft that you can scratch it with your fingernail. It will maintain this softness even at low temperatures. This is an excellent property if you are operating at low temperatures and do not want your material to become brittle at these low temperatures.

Indium has no *memory;* that is, if you shape it one way, it will not return to the original shape but will stay the way it was shaped. One area where this property has found application is in using pill terminations in stripline circuits. The case in which the stripline is put will have tolerances on its milled portions. For this reason, some of the cases could put extra pressure on the termination, while others may just barely make ground contact. If you make the depression for the pill intentionally deep and put indium in it, the indium will fill the void, conform to the pill shape, make good electrical contact, and remain in that shape, even if the pill is removed from the case. There are other similar applications where indium could perform as well. Indium can be added to other metals to make them:

- Harder;
- More fatigue resistant;
- More ductile (hammered thin without breaking);
- Easier to melt;
- Have higher heat conduction;
- Have higher electrical conduction;
- Easier to bond.

These properties make indium a very useful metal. The improvements mentioned when added to other metals make it an excellent choice for use as a solder for microwave circuits. A primary area where indium solders are used is when soldering to gold or silver. An indium-lead (In-Pb) solder will be an excellent choice over the conventional tin-lead (Sn-Pb) solder because tin alloys soldered against either gold or silver tend to scavenge the gold or silver. (Scavenging is the attack of molten solder on a metal surface, with the solder actually dissolving the base.) The choice of using indium-based solder would be an excellent one if, for example, you were using alumina substrate with chrome-gold metallization. This would ensure that the solder joints would be solid and reliable and exhibit high electrical conduction (something tin-lead solder probably would not do).

With all of the properties listed, indium and indium solders should be a strong consideration for use in microwave systems. Be aware, however, that the metal also exhibits some properties that may be less than favorable. Reports have been filed that indium solder may develop long-term problems if all intermetallic interfaces are not considered. These problems seem more evident in space applications. With this in mind, be sure to investigate all of the conditions, both short-term and long-term, before choosing any solder or metal to be used for your application.

A general description of indium is given here:

Symbol	In
Atomic weight (compared to carbon 12)	114.82
Specific gravity	7.3 (about 7½ times as heavy as water)
Color	Silvery white with brilliant metallic luster
Properties	Very soft material; stays soft at low temperatures; remains in shape when shaped; good conductor of heat and electricity

3.6.2 Tin

Most of the time, tin is used either in a tin-lead solder or a tin-lead plating. About the only time you hear of tin used alone is in tin cans, and even these, curiously enough, are not pure tin but a thin tin plating on an inexpensive metal. Today, many tin cans are not tin at all, and in the not-too-distant future any reference to a pure tin product may be a thing of the past.

Tin is another of the ancient metals that have been around for ages.

Tin articles were known to exist at least as early as 1400 B.C. in Egypt. Homer called tin *kassiteros* because he said it came from Cassiterides in the Atlantic. (The Cassiterides, or Tin Islands, were the British Isles.) The metal later had the name *stannum,* which resulted in its present symbol, Sn.

The only important ore of tin is cassiterite, or tinston, SnO_2, which is found in veins and streams, called alluvial deposits, mainly in Southeast Asia. (Bolivia and Africa also mine a certain amount of the metal.) Practically no tin is produced in the United States, which ranks first among users of this silvery-white metal.

As previously stated, the primary uses of tin in microwaves are in solder and for plating applications; the metal is excellent for both of these tasks. Remember, however, the precaution in the previous section about using tin-lead solder on gold.

A general description of tin is given here:

Symbol	Sn (from the Latin *stannum*)
Atomic weight (compared to carbon 12)	118.7
Specific gravity	7.29 (about 7½ times as heavy as water)
Color	Silvery white with a bluish tinge
Properties	Soft and malleable; melts at a little over twice the boiling point of water; at very low temperature crumbles into a gray powder; conducts heat and electricity reasonably well

3.6.3 Lead

Lead, like tin, is one of the metals that do not appear alone in microwave applications. You will recall how we referred to tin-lead solder and tin-lead plating. Similarly, indium-lead (InPb) solder is an excellent substitute solder when gold is involved. Regardless of the combination, it seems that it always appears with another element.

Use of lead has its origins back many centuries, as do many metals. A small lead statue in the British Museum dates to 3400 B.C., and lead is also mentioned in the Bible. In ancient days, lead was sometimes confused with tin, and for this reason, the Greeks used it very little. The Romans, however, used large quantities of what they called *plumbum.* Their main use was for water pipes. Actually, it is the word *"plumbum"* that resulted in our word *plumbing* and in the symbol Pb.

The microwave applications, as previously mentioned, are in solders and in plating. These uses are somewhat limited but provide excellent materials for joining or protective plating microwave materials and components.

A general description of lead is given here:

Symbol	Pb (from the Latin *plumbum*)
Atomic weight (compared to carbon 12)	207.21
Specific gravity	11.3 (more than 11 times as heavy as water)
Color	Bluish-gray
Properties	Soft and easily shaped; resists corrosion by sea water, air, and many chemicals; fairly low melting point; the worst conductor of electricity compared to all metals covered (although not a bad conductor)

3.6.4 Kovar and Invar

Two other "metals" should be mentioned here, as they also find applications in microwaves where thermal considerations are important. These are kovar and invar. These alloys—which they are, as opposed to pure metals—will be listed to be used as references later.

- *Kovar:* An iron-nickel-cobalt (Fe-Ni-Co) alloy with a coefficient of expansion similar to that of glass and silicon. Its thermal characteristics are similar to those of alumina substrates. It is used as a material for the mounting of alumina substrate to aluminum cases to compensate for differences in expansion, for headers, and in any glass-to-metal seals. (Notice that the three metals contained in this alloy are all in the ferrous metal group of the chart shown in Figure 3.1.) This metal can be difficult to machine.
- *Invar:* An alloy containing 63.8% iron, 36% nickel, and 0.2% carbon. It has a very low thermal coefficient of expansion and is used where thermal considerations are of prime importance and a minimum of machining is necessary. This material also can be difficult to machine.

3.7 Metal Applications

Before leaving the topic of metals in microwaves, we will present a review of the applications of each of the metals covered.

Aluminum: Used for cases for microstrip, stripline, and suspended substrates. Aluminum is also used as a backing for laminates by being bonded to the laminate and machined to be the baseplate for the case of a circuit. It can be cast to be a piece of waveguide with the skin depth plated with silver. This application makes a very efficient waveguide, which weighs much less than a waveguide made totally of copper or silver. Also, the cost is much less with a cast part that is internally plated.

Copper: Used primarily in microwaves as the metallic medium on laminates and substrates. It is used on the laminates because of its ease of attachment to the laminate, its excellent conductivity, and its ease of machining. The copper is used on hard substrates, as mentioned, to provide a surface to which to solder the fabricator. Without the copper layer, nothing would be solderable. Copper is also sometimes used in the fabrication of cavities and filters. Many times, a first model is made with pure copper and then other methods are used in the production of such a device. A problem with using only copper is that it becomes very heavy very fast. Many of the breadboard filters mentioned above would need a couple of people to pick them up. If you have a space application, this is not good. Remember that to operate in space, a component must be designed to weigh nothing, consume no power, and take up no space. Thus, weight would be very prohibited with a totally copper component.

Silver: Silver is used primarily in the waveguide areas of microwaves and higher frequencies. Because of its cost, the application discussed in Section 3.4 on waveguide plating is the one that has remained throughout the years. This metal is an excellent conductor and, as such, should be used whenever possible. You will also find some silver in a select group of solders. Remember, solder is not just meant to attach wires and components together; it also must be a good conductor of electricity.

Gold: The main application of gold in microwaves is in plating. The plating of cases for microwave circuits and plating of lines on stripline and microstrip are the primary areas where you will find gold used. The use of gold not only improves conduction but also protects such metals as copper from the oxidation process, which increases the surface resistance of the lines and the overall losses of the system. Gold is also used as a metal on hard substrates. You will recall from our discussions that gold will not adhere to ceramic materials. So there must be some buffer layer sputtered on the ceramic prior to the deposition of gold on the substrate. This, as we have said, is generally chrome.

Indium: This metal is primarily used in solder compounds. Its excellent conduction and low melting point make it excellent for this application. Indium is also used where a soft metal may be needed, as in the pill termination example presented in Section 3.6.1. Regardless of the other applications, indium will always be considered as the problem-solver in solders for microwave applications.

Tin: This is the main ingredient in the well-known tin-lead solders with which everyone is familiar. It is also used for a wide variety of plating applications. The tin-lead platings are used for many applications where the copper traces of a circuit must be protected from oxidation and the gold plating is not needed.

Lead: This metal is also used exclusively for solders. It is the other important part of the very common tin-lead (60/40) solder, familiar to all in electronics from the very first course in dc circuits and on up into the microwave regions. It causes some problems, however, as discussed, but still finds many applications in microwaves.

Kovar and Invar: These alloys are used as headers where there is a large discrepancy in the thermal coefficients of expansion of metals. One example of this is an aluminum case with a ceramic substrate attached to it. There is approximately a 4 : 1 difference in the coefficients of expansion of these materials. Any drastic change in temperature will crack the connections between these two units. With a kovar plate (header) between them, this difference is compensated. Remember, however, that both kovar and invar are difficult to machine. They should be used only when there is no other alternative.

3.8 Chapter Summary

This chapter has covered seven metals and two alloys that play an important part in microwaves. You will recall that the two properties necessary for usage of a metal in microwaves were electrical conductivity and machinability. With each of the metals covered, we stressed their conductivity, and each description listed as one of the properties that the metal was a soft metal. The only exceptions were the alloys, kovar and invar, which can be difficult to machine.

From the material covered in this chapter, you should be able to understand why metals are an integral part of the microwave circuits and systems in use today and those of the future.

4

Microwave Artwork

4.1 Introduction

The area of microwaves that has changed almost as much as the different types of material available is that of microwave artwork. That is, the methods used to generate the all-important negative or positive film used to actually etch the circuit board that the designer worked so hard to produce.

If you look at artwork for a microwave circuit, you can relate it to a pattern. If you are going to build a woodworking project that involves a series of elaborate cuts with a saw or lathe, you will need some sort of a pattern to do this. Similarly, if a shirt or blouse is to be made, a pattern must exist to let the seamstress know where to cut, where to sew, and where to add such pieces as cuffs and a collar. In each case, the success of the finished product depends on the accuracy of the pattern used to produce it. If the pattern is not precise, the finished product will not be the fine quality that is required. If the time is taken to make sure the pattern is accurate and exactly what is required, the final result will be a work of art. This is what a piece of microwave artwork does for a designer. It is that pattern of traces that will make the design function as it was intended. It shows the etching machine which material to remove and which to leave on the board as a transmission line. This is probably the most critical, and many times the most overlooked, portion of the microwave design process.

It is very nice to go through all the calculations of a new design and work with a computer to come up with the "ultimate" in a microwave circuit. This is a rewarding part of the process. However, if you think about it, it is also rewarding to see that same circuit in a case—tested, working, and ready to be placed into an overall system. What many people do not realize is that the main bridge between the "ultimate" design that comes off the computer and the final product for the system is the artwork that defines how this circuit will look. If you take shortcuts when making up the artwork, you will end up with a product that looks like you took shortcuts and probably cannot be used. This would be like designing a fast race car and saving money when you specified the tires so that the car would only run for a couple of miles before the tires wore out. A little more time spent doing the complete job would result in this car performing for the entire race. Similarly, more time spent on the "proper, accurate" artwork will ensure that the circuit will perform for the entire lifetime of the system it will work with.

Usually when fabricators and designers come together to discuss microwave circuits, the fabricators' major complaint is that the artwork supplied by the designer was not adequate to produce the circuit the designer requested. This complaint can be eliminated if the designer takes a little time and effort to know what is needed to produce an acceptable piece of artwork that will please even the fussiest of fabricators. This can happen by sitting down with the fabricator to learn about that particular fabricator's capabilities and making every effort to comply with those capabilities.

The goal, however, should not be simply to please a fabricator but to produce a piece of artwork that is the best it can be so that you can obtain a circuit that, if designed properly, will work as you expect it to. To this end, the microwave designer must realize that microwave artwork does not just happen because we want it to. There is a definite progression that must be followed: 1) an initial layout, 2) a final layout, and 3) the photo process. Each step will be discussed in this chapter, and many hints will be given to aid in producing the best piece of artwork for the particular circuit that has been designed.

Before getting into details, it should be stated that the artwork we are referring to is the final film that is actually used to expose and etch the microwave circuit. This may be a positive exposure, negative exposure, single-

Figure 4.1 Positive artwork.

Figure 4.2 Negative artwork.

sided, double-sided with registration between the two sides, or any other combination required for a particular application.

To clarify two terms used previously, we should look at positive and negative exposure. When we speak of a positive exposure and positive artwork, we are referring to a film that has all of the area that will be removed during the etching process as the light areas. The actual transmission lines that are to be left are the darkened areas. This is shown in Figure 4.1. This process, as the name implies, is a positive etch process with positive artwork and is used where close tolerance line width is required or metal etch resists (plating) like gold or tin-lead is preferred.

The negative process is the opposite of the positive process. The transmission lines that are to remain on the circuit board are light and the areas that are to be removed are dark on the film. This is shown in Figure 4.2.

Negative artwork is usually used for a print-and-etch process where plating and total metallization requirements for line widths are the tightest. If you look at the commercially available kits for etching PC boards, you will find that they use a ferric chloride etch and your artwork will have to be a negative process for this etching method. So it can be seen that the process of making and using artwork in microwave circuit applications is one that should be studied very carefully to ensure that you are using the best method for your particular application.

4.2 Initial Layout

Before many of the computer layout programs became available, this section probably would have been called "Microwave Drawings." As a matter of fact, the first two editions of this book called it that. This is where you put the basic transmission lines in the layout to see how they fit and interface with such areas as dc bias, grounds, and interfaces with other circuit boards

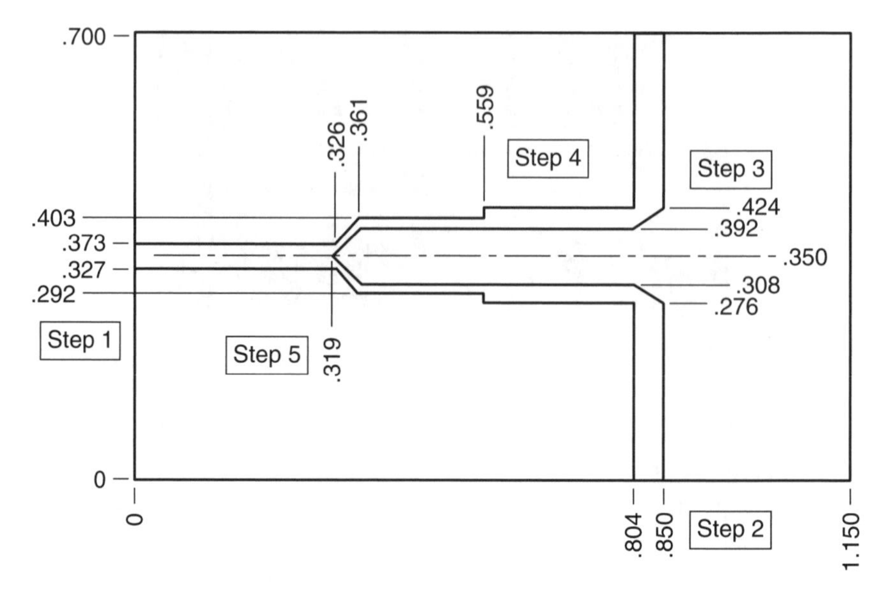

Figure 4.3 Power divider with dimensions.

or connectors. This drawing would also have a series of dimensions on it to allow the layout to be made with a certain degree of accuracy. As stated, this was done with a preliminary layout that would fit lines in the general area that they were allotted. An example of such a drawing, or initial layout, can be seen in Figure 4.3.

This is a two-way power divider drawing that has the circuit in the center and is completely dimensioned so that someone can make up a piece of artwork prior to etching the circuit. Today, all of these dimensions are put on a layout that is generated within a computer. When you see the layout on the screen, the transmission lines are shown at the width that was put into the computer, along with the correct lengths. At this point, you can see where each transmission line will be and its relationship to other lines.

The relationship of one transmission line to another is an important point to consider when making a microwave layout. If you look in any basic microwave text, there is usually a section on a component called a directional coupler. This is a device that has two transmission lines of a certain length to operate over a specified band of frequencies and that are placed close to each other so that the microwave energy can be "coupled" from one to the other. Such a circuit is shown in Figure 4.4. This is a good condition, since we are trying to couple energy from one transmission line to another as opposed to two transmission lines that should not couple energy but are placed too close together.

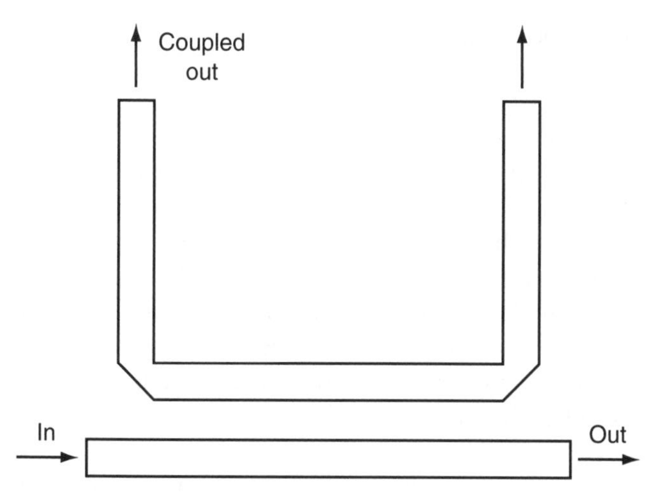

Figure 4.4 Directional coupler.

If, however, our circuit is not a directional coupler but simply transmission lines going from one point to another, we do *not* want to have energy coupled between these transmission lines. We want the energy to proceed from one point to another and not take any detours along the way. Each time that the energy is coupled to another line, there is a loss in that energy, and it is an expensive proposition to provide amplifiers in the microwave spectrum to boost this power level back up to what is required. In order to keep the power loss to a minimum, you should keep a couple of rules in mind when laying out a microwave circuit.

To understand the rules, we need to go back to a term previously used, that of *ground plane spacing*. Recall that this was designated by "b" and was one material thickness for a microstrip circuit and two material thicknesses for a stripline circuit. This is an important parameter when laying out microwave circuits.

The two rules to follow that will help you stay out of trouble when you make a microwave layout are:

1. Keep all transmission lines *at least* one ground plane spacing away from one another.
2. Keep all transmission lines *at least* one ground plane spacing away from the edge of the board and any grounds.

The term *at least* is important. For example, if you have a 0.030″ piece of microwave material that you are using for your circuit and you have a microstrip circuit to design, you should keep all transmission lines at least 0.030″

away from one another. As an addition to that rule, however, it is an even better practice to keep the transmission lines about 0.050″ from one another if at all possible in your layout. This is an excellent safety factor to throw in to ensure that you will not have an interaction between transmission lines.

Similarly, for stripline and a 0.030″ piece of material, the rule says that the lines should be 0.060″ from one another. As a safety factor, you should keep them in the neighborhood of 0.100″ from one another, if possible.

In the same sense, the lines should be kept 0.050″ and 0.100″ from the edge of the circuit board or from any ground plane for microstrip and stripline, respectively, for the example above. The dimensions from the edge of the board and from any grounds are there to keep the stray and fringing capacitances within the circuit to a minimum. If the transmission line is too close to the edge of the board or to a ground plane, you have a transmission line in close proximity to a ground, which makes an excellent capacitor.

One of the largest enemies of microwave energy is capacitance. Since microwaves are so high in frequency, the reactance (RF ohmic value) of a capacitor becomes very small at these frequencies and is basically a short circuit to ground for the energy. This, obviously, is not what is needed or desired for the proper operation of a microwave circuit. Thus, adherence to the two rules stated above is vitally important when laying out any microwave circuit.

Another factor to consider when making an initial layout of a microwave circuit is how the transmission lines will be bent to fit into the small area allotted to them. Some of this was discussed in Chapter 1, when we meandered a transmission line to have it fit onto a certain size circuit board. In order to do this meandering process, it is necessary to have corners of some type. The topic of microwave bends has been investigated so much that there is very little that can be said about bends, or corners, that has not been written about many times over. The one thing to keep in mind when considering bending a transmission line is that microwave energy does *not* like to make square corners. Square corners may be exactly what you need if you are in a military parade and want to look sharp, but square corners will only get you in trouble in a microwave circuit.

The main problem with square corners is the reflections that occur when the microwave energy encounters such a corner. If you try to shine a flashlight down a square-cornered pipe, you know what will happen. The light will hit the end of the input pipe and reflect back toward your flashlight. Very little light will ever get around the corner. It is the same for microwave energy and a square corner. Very little energy gets around the corner. Most of it is reflected and lost.

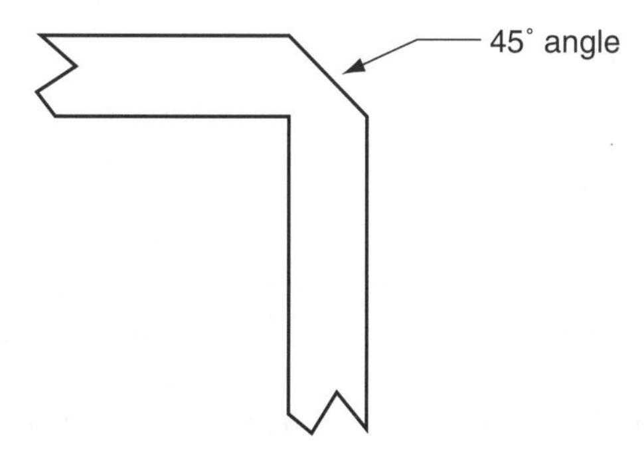

Figure 4.5 Mitered corner.

As we said, there are a variety of ways to get the microwave energy from one point to another around a corner. The method that seems to have found the most favor is a straightforward miter corner. This is the corner that is a 45° angle on the corner and is shown in Figure 4.5.

It can be seen that this is an efficient method of getting the energy around the corner. It also goes back to the idea of optics where the angle of reflection equals the angle of incidence that you studied in high school physics. The similarity between optics (light) and microwave energy is not coincidental. The two forms of energy have a lot in common and react many times in the same way. Thus, the mitered corner is one that is used in many microwave layouts.

As a means of improving the transition of microwave energy around a corner, Figure 4.6 shows a method that lowers the loss around the corner from meandered transmission lines.

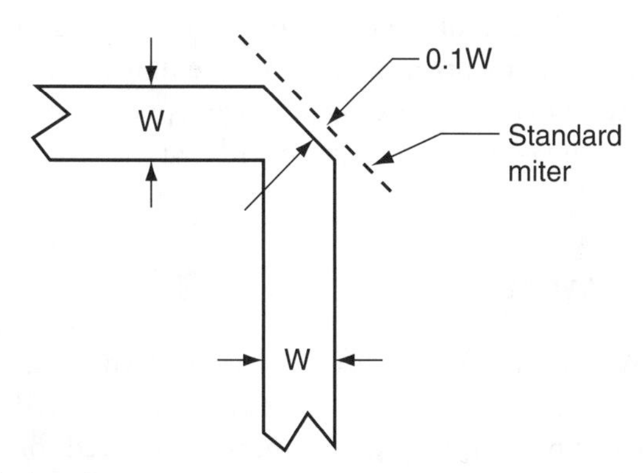

Figure 4.6 Improved mitered corner.

As you can see, the corner is still mitered but the sloping edge is a bit different than that of Figure 4.5. The actual slope is recessed by 0.1W, which is one-tenth of the width of the transmission line being bent. This has been found to be an excellent corner, especially for a circuit board that has a series of corners, such as a long delay line. The less loss you can have in a simple corner, the better the performance of the overall circuit. Each individual corner may look like it has an insignificant loss, but when you add them all up, it comes out to be very significant and perhaps a loss that will be the difference between a working circuit and a paperweight for your desk.

One final area to cover before we proceed to the final layout section of this chapter is the layout of transitions from microstrip and stripline circuits to connectors. This is a transition area that must be addressed in order to eliminate any mismatches between the circuit board and connectors. These mismatches will produce losses that will hinder the performance of the circuit. This is an area that is ignored many times by designers and people who do layout work. They will reason that if we have a 50-ohm transmission line at the output of the circuit board going to a 50-ohm connector, everything should work just fine. What they do not realize is that the 50-ohm line on the circuit board is in stripline or microstrip, and the connector is a coaxial configuration. Regardless of whether the connector is a Type-N, SMA, TNC, or any other, it is still a different medium. The interface between these two mediums is as critical as which mediums are chosen in the first place.

Figure 4.7 shows a general rule of thumb that can be used for stripline and microstrip circuits to aid in matching your transmission line inputs/outputs to the connector, flexible ribbon, movable tab, or any other transition method used. This is a matching pad for the input/output transmission lines going to or coming from a coaxial connector. The dimensions state that there are a W (width of the input/output line), 2W (width of the matching pad), and 1.5W (depth of the matching pad), that must be considered. There may be some adjustments necessary to these dimensions for specific applications, but this drawing is an excellent starting point and should be used for all transitions when laying out a circuit board that interfaces with a connector.

4.3 Final Layout

Now that we have a general idea of what our layout will be and have taken certain precautions in the layout process, we can proceed to make up a final layout that will be used to produce the artwork used for the circuit etching process, which will be addressed in Chapter 5.

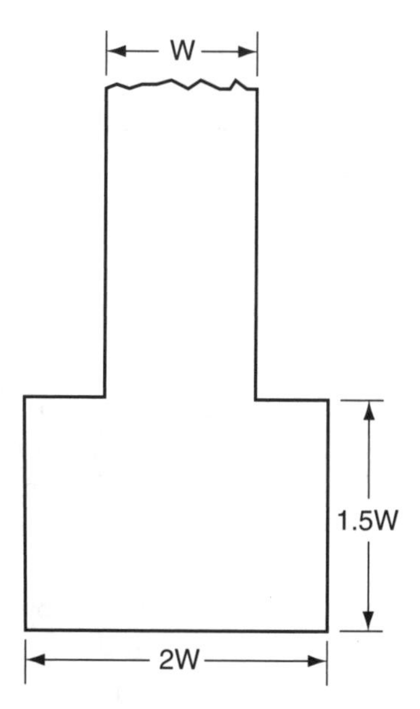

Figure 4.7 Transmission line transition.

Through the years, there have been many methods used to produce a microwave circuit layout so that artwork can be made and the circuit etched from that artwork. In the early days, there was printed circuit (PC) tape used to lay out a particular circuit. The tapes were standard widths used to reproduce transmission line widths for microwave circuits. The first stripline circuits were produced by placing the tape on Mylar in the layout representative of the required circuit. The circuits were usually laid out 10 to 20 times larger than required and then photographically reduced to obtain the needed accuracies. This tape was a *pressure-sensitive graphic tape* or *precision-slit PC network tape.* It was generally a black tape but may also have been a photographically opaque red transparent tape. The tape's primary requirement was that it block the photographic light when the artwork was exposed.

In some locations, the PC tape version of laying out artwork may still be used. Those who cannot afford the higher priced computer-generated artwork programs may rely on some of the more basic methods, which still work very well when they are the only methods available. In most cases where the PC tape version is used, the accuracy and smoothness of the line edges are insufficient for most of the circuits required for today's microwave technology. With this in mind, other forms of microwave circuit layout are needed.

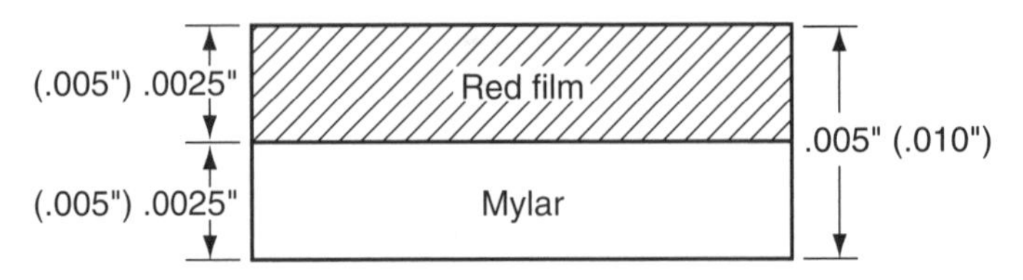

Figure 4.8 Rubylith construction.

Another method used for early microwave circuit layout, and still in use today, is the rubylith. The term *rubylith* cannot be found in any dictionary, electronic or otherwise. It is a term used by the Ulano Company as a trade name for a photographic masking film. The name has been so widely used over the years that it has become *the* accepted term for this type of film throughout the industry.

The film consists of a Mylar sheet covered by a peelable layer of soft transparent red material. The construction is shown in Figure 4.8.

The dimensions shown in Figure 4.8 are typical dimensions with the ones in parentheses being for a popular thickness material. It can be seen that the thickness of the Mylar and the red material are the same in both cases. This gives the film a good Mylar base while producing a well-defined edge on the red film when cut.

The edge on the lines and the mitered corner of a rubylith have been cut. A particular transmission line was cut by hand with a knife. The accuracy of the transmission line is not the best because of the possibility of human error when the cuts are made. For many applications today, the cuts are automatically made from computer-generated programs with a high degree of accuracy.

Rubyliths being cut from computer-generated programs is the main thrust of the technology today in generating microwave artwork. There are a variety of programs that combine the tasks of circuit design, circuit board layout, and artwork generation. This is a convenient way of producing a microwave circuit that stands a good chance of working as you predicted.

There are some areas of the computer-generated artwork scheme that should be discussed before we talk about specific programs available as of this printing. One of the first areas—how much undercut is expected from the etching process—was referred to generally when we previously talked about the interface between the designer and fabricator. In other words, what is the final dimension of the copper trace after etching compared with the dimension that is on the drawings and artwork? It is not very nice to design a circuit that, for example, requires 0.046″ transmission lines for proper operation and then find out that things do not work very well when you actually build the circuit. If you measure the width of the final etched transmission lines, you may find that they are only 0.044″ wide instead of 0.046″. Where did the other 0.002″ go? Well, this is the amount that the etching process undercut as the process went through its normal cycle. (Many times it is not this severe and other times it may be more severe than 0.002″. The point is that you should be aware of just what the undercutting will be so that you can compensate for it.)

If you really need to have 0.046″ lines, you should find out how much undercutting will occur and then you can compensate for that in your artwork. In the case above, you should lay out your artwork so that all 0.046″ transmission lines are actually laid out to be 0.048″ and your circuit will stand a much better chance of working as you designed it.

Another area to keep close watch on is when you have coupled transmission lines. There will be a certain spacing required between the transmission lines in order for the coupler to operate properly. Recall our previous discussions on coupled transmission lines and directional couplers. If you require a spacing of 0.020″, for example, and you know that there may be a 0.002″ undercutting taking place, you should lay out the artwork so that the actual spacing on the artwork is 0.018″ instead of 0.020″ as required. If the etching process has this 0.002″ undercutting, the circuit will come out just perfect.

The two examples presented above are similar to throwing a football or baseball through a suspended tire that is swinging back and forth. If you aim right at the tire itself, you will probably miss it. If you "lead" the tire and anticipate its ultimate location, you will throw the ball through the opening instead of missing it completely or bouncing the ball off of the tire. There is a compensation used in order to ensure success. The same is true for microwave artwork. If you know the limitations and tolerances on the etching process you will be using, you can compensate for that process and have the transmission lines in the final circuit be exactly what you wish them to be.

We stated that there are a variety of programs that combine the design, layout, and production of artwork. We also said that these are very convenient for designers to use once they have factored in the line width and spacing corrections discussed above. The time and energy savings for such programs can be realized when you consider the procedure that was followed many years ago to produce a finished piece of artwork that would work as predicted. A drawing was made that was the general layout of the circuit, which was many times (10 to 20 times) the finished size of the circuit. This drawing was dimensioned, usually starting at a datum point in the lower left-hand corner of the drawing; the drawing was then used to cut a rubylith or produce a photographic picture of the circuit. Many times, the PC tape was laid out 10 to 20 times and the layout was photographically shot and reduced to obtain the final size and accuracy. This artwork was then used to etch a circuit board and any corrections had to go through the same process to obtain the finished circuit.

As we stated, this can be costly and time-consuming. The computer artwork generation programs are much more efficient and much faster than previous methods. You should be cautioned, though, that the quality of the artwork is only as good as the information put into the program. It comes down to the same old saying, "Garbage in, garbage out." So remember that the programs require you to know what you are doing to produce the quality piece of artwork necessary.

One of the programs available for microwave artwork layout is Microwave Office™. We will go through this program first and explain its features and how they relate to the typical designer's requirements. After that, we will present other microwave artwork programs in the same way.

The Microwave Office program combines many features that are valuable for both the novice and experienced designer. This program will produce the layout of your circuit and send it to the appropriate equipment to have the artwork made up for a final circuit.

One factor that is useful in obtaining a final design that will function as predicted is the program's real-time tuning capability. The designer can select components and sketch them in while the program presents the output changes to show the circuit performance.

Another feature of Microwave Office is a three-dimension photo realistic view that is available. If you are doing an audio amplifier design, for example, you will use a transistor (or transistors), resistors, capacitors, and some method to get a signal in and out of the amplifier. When you build this circuit, you have the components readily available to you and you either put them on a prototype board or solder them together without a board. The point is that you can see the components and where each can go and how much room they will occupy. You can also see if one component is going to get in the way of another one. With microwave circuits, this is not possible. The only thing that came even close to this was a material developed over 20 years ago that had a very low peel strength. This material could be used to make up a breadboard by cutting circuits out with a knife and peeling away the excess copper. With this material, you could build a breadboard similar to the method we referred to for the audio amplifier. This was

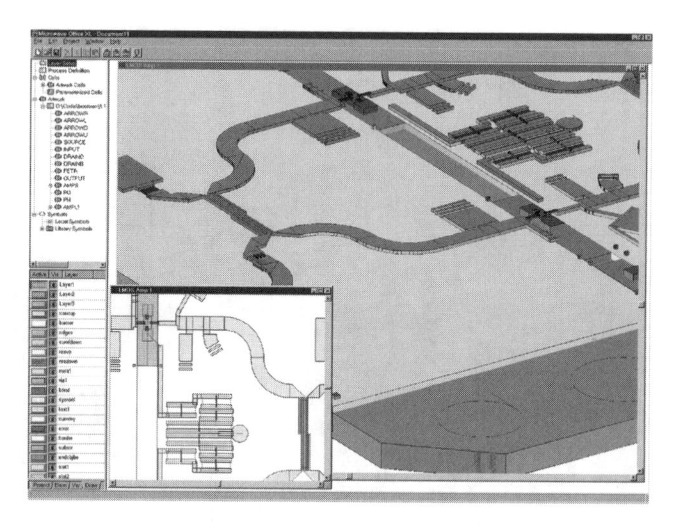

Figure 4.9 3-D view of transmission line layout.

somewhat of a passing fancy and did not go far. So we still need to build the circuit and see where the problems lie.

With the three-dimension photo realistic view capability, a 3-D view of the transmission lines and where they will encounter other transmission lines is presented to the designer. Such a circuit is shown in Figure 4.9. This capability allows a designer to look at the final design from any angle to see if there is a clearance problem or a connection problem within the final circuit.

One problem area with some of the early computer programs for circuit layout in microwaves is that of tees, bends, and steps in transmission lines. It was usually rather difficult to accurately represent these discontinuities so that the output you observed on the computer screen was representative of the output you actually observed with the finished circuit. Usually the problem was not caused by the straightforward transmission line but by any time that the transmission line had to change direction or width.

The Microwave Office approach is to create cells that are classified as parameterless. This allows the models of these discontinuities to adapt based on what they are connected to. This feature does not put the typical constraint on the set discontinuities but allows them to be more flexible to the actual environment they are in. Figure 4.10 shows an example of these cells and shows the tees, steps, and bends referred to previously.

The final section covered for this layout program is how we get a piece of artwork from the computer layout we just produced. After all, this is a chapter called "Microwave Artwork." To produce this artwork, the program

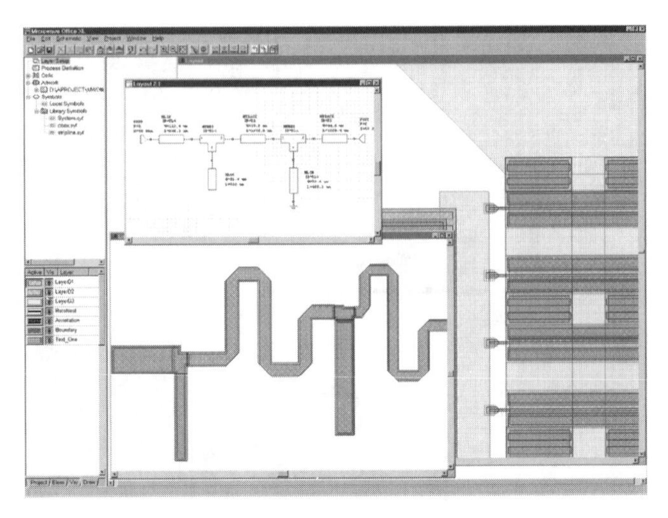

Figure 4.10 Circuit layout cells.

must be compatible with a plotting system of some kind. Usually this is a Gerber plotting system that will produce very accurate artwork that is used in many applications for microwave circuits today and has been used for many years. This program has the capability of interfacing with the Gerber systems as well as DXF and HPGL. (These are both plotting systems that are available to produce the final artwork.) Designers should be able to obtain a very accurate plot of their circuit and artwork that will allow the circuit to operate as they intended.

Now is a good time to discuss such areas as the plotting systems that need to interface with these new sophisticated layout programs. As we said, the Gerber system has been used for many years and is reliable and accurate. There are those, however, who have sought to replace the Gerber with other systems. Articles and studies both support the theory of Gerber and stress that it is outdated and should be replaced. We will not side with either theory but only provide information on both sides and let you make your own choice.

Before getting into this discussion, we will define a term that will be referred to throughout this section and may be seen in other chapters. This is the CAD-to-CAM data transfer. That is, the computer-aided-design to computer-aided-manufacturing data transfer. We are all familiar with the process of data transfer. It gets very frustrating when you do your required task on a computer and then cannot get the information to a printer or plotter. This simple example should bring the importance of the proper plotting system, or standard, to light very quickly. After designing the circuit and having the layout put together properly, the next step should be to have a piece of artwork be the exact duplicate of that layout so that the circuit can be fabricated (manufactured) properly. Thus, the data transfer must be fast and efficient, which means that the plotting system and computer must be able to converse with one another. To sum up the importance of this CAD-to-CAM data transfer and the standard that governs this transfer, we can sum up what such a standard should accomplish by stating that the purpose of such a standard is:

> To increase efficiency, minimize errors, and increase throughput. This will yield higher productivity and, thus, lower costs.

If you look at this statement, you realize that any standard should conform to this definition.

The standard used must be supported by both the CAD and CAM vendors or the system will obviously not work. This is similar to the idea that you may come up with this fantastic telephone that will dial from your voice command and transfer to a fax machine or e-mail simply by thinking that instruction. This would be quite a telephone but probably could not be used because no one else would have a telephone or other equipment that would be able to converse with it. This is the idea behind getting the CAD and CAM vendors on the same page.

The Gerber standard has been around for many years. Its greatest advantage is that it is a very simple system and many companies are set up to handle it at this time. It also has taken more than 10 years for the industry to completely understand it and be able to incorporate it to its highest efficiency level. Another argument discussed whenever a new standard is presented for consideration to replace Gerber is that changing a standard takes many steps and revisions before approval is granted. This, as many of you know, translates into time and is something that most manufacturers are not willing to give. They have a working standard now and would prefer to keep working with it. Most of them, however, would consider some new standard *after* the bugs were worked out. Debugging a new standard takes time, and time, as the old saying goes, is money. And, after all, that is really why a company is in business, to make money, or as economic professors say, earn profit.

The other side of the coin produces people who argue that today's designs are more complex and that we need to replace Gerber with a more sophisticated standard that will more accurately reproduce these complex designs. This may well be a valid argument.

Table 4.1

IPC-2511	Generic Requirements for Implementation of Product Manufacture, Description Data, and Transfer Methodology.
IPC-2512	Sectional Requirements for Implementation of Administrative Methods for Manufacturing Data Description.
IPC-2513	Sectional Requirements for Implementation of Drawing Methods for Manufacturing Data Description.
IPC-2514	Sectional Requirements for Implementation of Printed Board Fabrication Data Description.
IPC-2515	Sectional Requirements for Implementation of Bare Board Product Electrical Testing Data Description.
IPC-2516	Sectional Requirement for Implementation of Assembled Board Product Manufacturing Date Description.
IPC-2517	Sectional Requirements for Implementation of Assembly In-Circuit Testing Data Description.
IPC-2518	Sectional Requirements for Implementation of Part List Product Data Description.

One of the standards that has been proposed is from the IPC and is called GenCAM (generic computer-aided manufacturing). It is designated as IPC-2511, although there are many supplemental documents that go along with the IPC-2511 document. It has the standard ASCII (actually it is USASCII or United States of America Standard Code for Information Interchange). It consists of 20 sections that convey the design requirements and manufacturing details for the standard. Each section is independent but their relationship is important to the user. IPC-2511 acts as the identification for a set of standards that provide the generic explanation of the GenCAM format. The complete list of documents that pertain to GenCAM are shown in Table 4.1.

It can be seen from our general presentation of Gerber vs. non-Gerber that there are many hurdles to get over before any standard can completely replace the Gerber standard. That is not to say that it cannot be done. It probably is inevitable that in the future most of the Gerber systems will be converted to some other standard, which is a good thing since the worst thing that can happen to any industry is to stagnate and resist all change.

With the Gerber issue addressed to some degree, it is now time to get back to the topic we originally began to discuss—computer-aided design and manufacturing programs that are available. One such program is the MMICAD Schematic/Layout package. This program features a top-down design capability while still supporting the bottom-up design procedure. A designer can start with either a circuit envelope or the individual components.

For circuits of moderate complexity, the user can generate the circuit schematic automatically from either a layout or the circuit netlist. The MMICAD layout is based on the CALMA GDS II stream format. CALMA is an interactive graphic system that can be used to lay out microwave circuits. By placing x,y coordinates from a drawing into the system (or using a file from the program), we can obtain a layout of the circuit in the form of a coordinate printout, pen plot of the circuit, and the complete circuit on tape that can be given to a plotter to make artwork. The system consists of a display terminal with a keyboard, two displays (one with the layout and one with the numerical information), and a tablet with an electronic tip for entering data points manually, the CPU (central processing unit), a tape reader, and a printer. Many companies used the CALMA system to lay out circuits before computer-based systems were designed. Many companies still rely on the CALMA to do their layouts today.

The MMICAD Schematic/Layout also has outputs available for Autocad, DXF, HPGL, IGES, Excellon NC, and Gerber formats. As we have previously seen, this output versatility is an important part of tailoring a layout system to the industry that must eventually make up the artwork for the circuit that has been designed.

A third program available for circuit layout to produce artwork for microwave circuits is =LAYOUT=, which can work in conjunction with a schematic entry program called =SCHEMAX=. Both are part of the overall Genesys software that is a design, schematic, and layout program used for many microwave circuits.

As mentioned, the =LAYOUT= portion is what we are concerned with in this chapter. To obtain the necessary layout, designers can use the =SCHEMAX= files from the overall program, manipulate the layout as needed for their application, and then the =LAYOUT= portion plots the artwork or generates an optimized Gerber file that can be sent to a printed circuit board house for fabrication.

=LAYOUT= contains a large library of footprints (land patterns) for active and passive components based on the IPC surface mount and land pattern standards (IPC SM-782). There is also a library derived from published manufacturer's data.

As mentioned, the program generates a Gerber file. It also generates an Excellon drill list where needed for use by a PC board manufacturer. The Gerber files are also ready for use with printed wire board milling machines.

Many times, these files can be fed to commercially available machines that will cut the traces on the board as well as trim it to size. These machines will be covered in the next chapter.

We have concentrated on the idea that a computer design is to be transferred to some sort of plotter to make up the artwork. This is a true statement but needs to be explained further. We have left an impression that the only way that the artwork can be produced is by having some cutting device cut a rubylith or other similar material to make up the artwork. In reality, there is another method for making very accurate artwork. That method is using a laser to shoot the artwork. As stated, this is a very accurate method for obtaining accurate line widths for your transmission lines. The systems use a raster of dots to produce the required areas on the artwork. The smaller the dots, the more accurate the artwork will be. This is just one area where lasers are being used for systems where a high degree of accuracy is required.

4.4 Chapter Summary

There are a variety of methods available to generate the microwave artwork that is needed for very precise microwave circuits. The methods available can range from PC tape for very general circuits that are not required to have a high degree of accuracy, to the rubylith being cut by hand with a knife or from a computer program, to the artwork being generated within a computer program and placed in the appropriate file to have the final artwork produced either by a cutting method or laser system. The methods are wide and varied, but all are aimed at the same result—to take that one of a kind, fantastic microwave design and make it a reality. This has been the purpose of this chapter.

It can also be seen that the opening statement of this chapter is very true. That statement said that the area that has changed almost as much as microwave materials is that of microwave artwork. This is an excellent connection since any changes in microwave material will, in many ways, affect the way they are fabricated. It should be very clear at this point that microwave artwork is just as special as the microwave materials that it is placed on to produce the circuits. Obviously, we are not talking about a positive or negative that is only there to make connections between components or to carry current from one point to another. It is a vital part of the final working circuit for microwaves.

5

Etching and Plating Techniques

5.1 Introduction

In the previous chapter, we said that the area that has changed almost as much as the types of microwave materials is that of microwave artwork. The use of computers to generate artwork has made a tremendous difference. Similarly, as the materials change and new ones are manufactured, the method of etching copper from them, the plating methods used, and the plated-through-hole fabrication techniques must keep up with the new materials. The new materials will react much differently with the chemicals previously used to produce the final microwave circuits. There was a time when the only equipment necessary to etch a circuit board was some ferric chloride, a pan to put the ferric chloride in, and some source of heat to accelerate the etching process. There are still some commercial products you can purchase that contain equipment very similar to the list just presented.

The process of *accurately* etching microwave circuits requires much more than ferric chloride and a heated pan. The highly sophisticated process results in very accurate width lines that will perform as the designer had intended.

Similarly, there was also a time when manufacturers would put on their data sheets that it was not possible to plate-through-holes on Teflon-based materials. That sounds ridiculous today, but it used to be a reality and indicates just how things can change. Designers and fabricators who do not keep up with these changes are going to be left behind.

Even though the processes are much more complex today, there are still some basic steps that should be performed for every circuit board that is to be etched or processed in any way. There is one basic, and important, step that should be considered when working with microwave circuit boards in any capacity. That step is:

- Cleaning the material *thoroughly*.

The most important word here is *thoroughly*. This is probably the Number One area with which anyone who has anything to do with etching or plating a microwave circuit board should be concerned. If there are any contaminates at all on the circuit board, the process will not be able to be performed efficiently or to the best standards of the equipment being used. Make this your top priority when working with microwave circuit boards.

In this book's previous editions, this chapter was called Etching Techniques. You will note that this chapter now covers etching and plating techniques. It was felt that the plating and etching process closely follow one another and thus should be together. We will cover both processes in that order, etching followed by plating.

5.2 Etching Techniques

With all of the tips and suggestions presented in the previous artwork chapter, you are now ready to etch that prize circuit and see how it is going to perform. To get started, it is important to understand some basic steps that should be followed when etching a microwave printed circuit board. When you observe these steps, you will have a good chance of producing the circuit board accurately and precisely so that it will perform in the proper way. The steps to follow are:

1. Clean the material thoroughly (just as we discussed earlier).
2. Carefully apply the proper photo etch.
3. Place the artwork on the material and expose it to identify the circuit area to be etched.
4. Pay careful attention to the actual etching process.

5.2.1 Cleaning

As previously mentioned, the cleaning of any microwave material prior to either etching or plating is of utmost importance and should be completed *thoroughly*. This is the same as when you want to paint your house. You will note that the manufacturer tells you that for best results, the surface to be painted should be clean and free of any oil or contaminate. If you do not clean the surface before you paint, you end up painting over a base of dirt and it will not last. The same is true for etching a microwave circuit board. When etching such a circuit board, it is necessary to put photoresist on the board to define which areas will be etched and which will not. If the resist does not have a good clean surface to adhere to, these areas will not be well defined and your transmission lines will be very sloppy in regards to both width and length.

Another way to look at this is to relate it to trying to put a piece of masking tape on a window in order to paint the frame more quickly. If the tape falls on the ground and gets dirt on it, it will not stick very well to the window and your job will be much more difficult than if the tape is completely clean and goes on the window tightly so that paint cannot get underneath it. With all of these examples, you can see that the cleaning process cannot be stressed enough. If there is any doubt that your surface is clean, clean it again to make sure.

To be certain that you are using the proper chemicals to clean the particular material you are etching, you should consult the manufacturer's recommendations. Knowing the *proper* cleaning method for your particular application, such as which commercially available chemical copper cleaners to wipe on the board or dip it in and then rinse and dry it, will greatly enhance your chances of producing an accurate and usable circuit board. We will not recommend specific methods of cleaning because there are variations that could cause you more problems than we are attempting to solve. Suffice it to say that all cleaning processes consist of three basic operations: (1) solvent bath, (2) rinse, and (3) baking (or drying). The number of steps may vary; more than one bath or one rinse may be involved, as is the case when hard substrates (alumina, for example) are prepared for metallization by a sputtering process. This requires much more than the basic cleaning process previously mentioned.

To understand the steps of a cleaning process, it helps to know what is being cleaned in the first place. There are two basic types of contaminates that must be removed in order to achieve a good etch: oil-based soils and water-soluble soils. Oil-based soils can be removed by vapor degreasing or using an ultrasonic cleaning process that basically shakes the contaminates off of the material. Water-soluble soils are removed by using deionized water or high-purity alcohol.

The deionized water, or DI H_2O, is a water that has been purified by removal of ionizable materials. Ionized materials are those that have electrons easily removed from molecules. This removal causes ions to be formed and an increase in free electrons that will reduce the electrical resistance of the material. For a thorough cleaning process, these materials should be removed from the cleaning solutions. Thus, deionized water is a primary cleaning product that should be available.

In order to fully understand the cleaning of a circuit board before etching it and producing a usable finished board, one point should be presented and emphasized greatly: do *not* touch the substrate or laminate with your hands. Use nonmetallic tweezers or rubber finger cots. Your fingers can leave a film of oil or dirt particles on the material that defeats everything you are trying to accomplish with the cleaning process. It can be rather discouraging to clean your circuit board, pick it up with your hands to apply photoresist, expose the board, etch the board the proper way, and then find out that there are discontinuities in your transmission lines that look exactly like the fingerprint of your right thumb. When handling laminates or substrates, use the proper precautions and save time and trouble.

The process of cleaning circuit boards is one that should not be taken lightly. It should be apparent by this point just how important a good cleaning is to prepare the circuits for etching.

As a review of the cleaning process, we will summarize the Do's and Don'ts as they pertain specifically to pre-etch cases—that is, how to clean a metallized substrate or copper-clad laminate prior to etching.

1. *Do not* touch a substrate or copper-clad laminate with your fingers. Use a set of nonmetal tweezers or finger cots.
2. *Do* use a liquid cleaning solvent. Avoid abrasive cleansers because they may scratch the metallic surface. These fine-grained scratches will cause problems with your microwave circuit.
3. *Do* use vapor degreasing or an ultrasonic cleaning for oil-based soils that may be present on the surface. Use deionized water or high-purity alcohol for water-soluble soils.
4. *Always* end the cleaning process with a baking step that removes any films left from previous cleaning operations.

If these steps are followed, you should eliminate many of the problems that can arise during the etching process. Remember that the premetallized and pre-etch cleaning processes presented are only examples of what can be done to clean the substrate or laminate. Knowing what you are trying to clean and the degree of cleanliness required will dictate what steps your particular cleaning process will have. It is a good idea to adapt any processes to your particular application and not memorize other processes that may not apply to you. This is why we offered a general presentation of the cleaning process with some suggestions.

5.2.2 Photoresists

The process used for etching metallized substrates or copper-clad laminates is called *photolithography* (or photoetch). This process uses photographic techniques and materials, which are photosensitive polymers (or emulsions) called photoresists. The International Society for Hybrid Microelectronics' (ISHM) glossary of terms defines a photoresist as "a photosensitive plastic coating material that, when exposed to ultraviolet light, becomes hardened and is resistant to etching solutions." This definition holds true for both types of resists: positive and negative. This may be a bit difficult to grasp if you think of a resist put on a board, artwork placed on it, and the resist exposed to ultraviolet light. When a negative piece of artwork is exposed, the definition makes sense because the circuit you want is exposed and the resist hardens and resists etching. If, however, you use positive artwork, you would expose everything you do not want and the circuit would be etched away.

This would be the case if you only took into account that you apply resist, place the artwork on the material, expose it to ultraviolet light, and etch. There is, however, one important additional step that is not included in the process listed above: the step that develops the resist. This step is the distinguishing factor between the processes. With this distinction in mind, the following statements can be made concerning photoresists:

> *Negative*—An emulsion that becomes insoluble when exposed to ultraviolet light. This makes the pattern you desire hard and resists the etching chemicals.

(Figure 5.1 shows negative artwork. You can see how the light will expose the pattern you desire and not the areas to be removed. Thus, the result on the material is exactly the opposite of the artwork: a *negative* image.)

Figure 5.1 Negative artwork.

Figure 5.2 Positive artwork.

> *Positive*—An emulsion that is soluble in developing solvents. This resist duplicates the pattern you desire, and thus we have the name *positive.*

(Figure 5.2 shows positive artwork. All areas to be etched away will be exposed to the ultraviolet light. The resist in these areas must be washed away by the developing process so that the only area with resist left on it is what you want for your circuit.)

To understand further how the positive and negative processors work, refer to Figure 5.3. In both cases, the substrate has a metallization on it. It may be ½ oz or 1 oz copper, a chrome-gold metallization, or a chrome-copper-gold combination. Regardless of its composition, there is a substrate with metal on it. The photoresist (positive or negative) is then applied to the metal (methods will be discussed later in this section). The artwork is hence placed on top of the resist-treated substrate, and ultraviolet light is projected down on the entire setup. In Figure 5.3(a), the light exposes only the desired trace. Following the developing and etching process, the only thing left is that trace.

Conversely, Figure 5.3(b) shows that the ultraviolet light exposes everything *except* the desired trace. This allows all of this exposed surface to be washed clear of photoresist in these areas and leaves it only over the desired trace. Once this process is completed and the etching is done, the result is shown as the same trace as in the negative process. We arrive at the same end by using two different processes: in one, the resist stands up to developing and etching (negative), while in the other the resist is rinsed away where it is not needed (positive).

The decision as to which resist to use is comparable to the decision as to which material to use. There is no *best* resist, but there are applications where one may give better results than the other. Generally, if you have narrow lines to etch or have a narrow gap between lines, it is best to use a

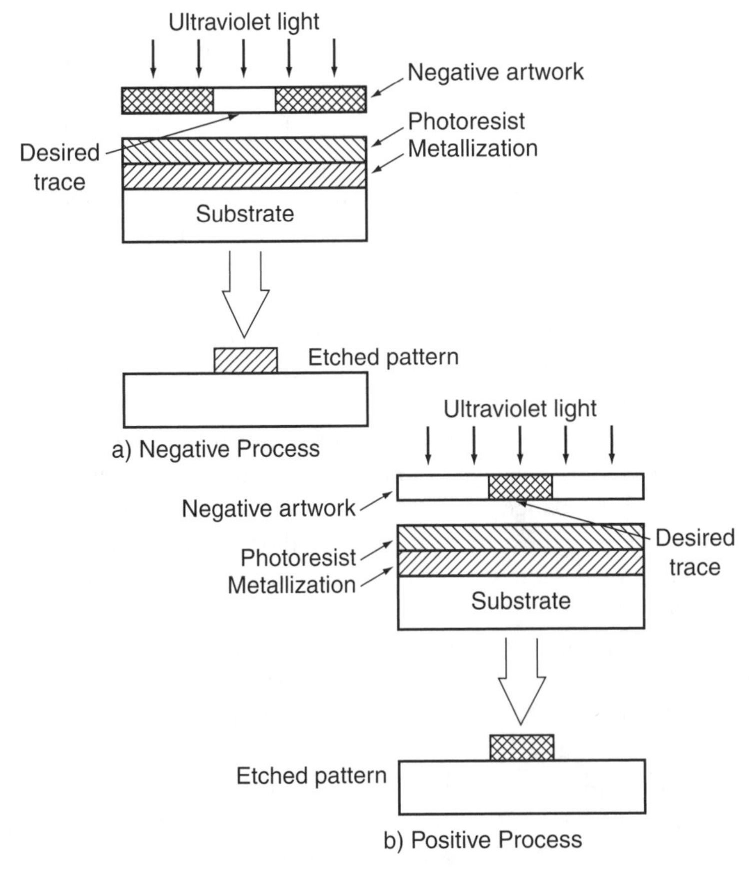

Figure 5.3 Positive and negative processes.

positive resist and, of course, positive artwork. This is because you can get a much sharper edge on a line or a gap that may be exposed to a higher degree.

Negative resist also has many applications. Up until the past few years, the largest majority of microwave circuits were etched using a negative process. Positive processes were used only in special cases. This has changed today, however, and the industry now seems to be favoring the positive scheme. The ultimate decision still should depend on your particular application and what is best for your circuit. Do not use a process just because the rest of the industry does.

There are various ways to apply resist to the circuit board to be etched: by *immersion*, by *spraying*, or by *spin coating*.

The immersion technique consists of dropping the entire substrate into a resist bath in order to completely coat the unit. This can cause problems with woven PTFE/glass material if the resist gets on the laminate and "wicks" into the material a certain distance. This disturbs the dielectric properties of the material on the outside edges, which destroys its uniformity. (Uniformity of material is necessary for proper microwave operation, as was stated in Chapter 2 on microwave substrates and laminates.) This process, therefore, is not a highly recommended one, although it is used in cases where it is the only method available.

In the spraying technique, the substrate or laminate is sprayed with the resist, using either a spray can or other spraying system. This process is much more acceptable because it avoids the edge of the material and also controls the thickness of resist much better than immersion in a bath. Even though the control of coverage is much better than in the immersion method, there still is a danger of making the resist too thick in one area or too thin in another area if a manual spray method is used. An automatic spraying system provides a much better coverage over the entire substrate or laminate surface.

The process of resist application used most widely throughout the microwave industry is *spin coating*. In this process, the substrate (or copper-clad laminate) is covered with the appropriate resist, and the excess is removed by spinning the substrate at a high rpm rate. The material is covered with a uniform coating of photoresist following this spinning, and the coating is then air dried and prebaked at a temperature that allows the solvent to evaporate without degrading the photosensitive properties of the resist.

To illustrate the photoresist technique, we describe two processes: one for copper-clad laminates and the other for a metallized alumina substrate. Note how they follow the basic description presented above.

Process 1: Copper-Clad Laminates

- Following a thorough cleaning as described earlier (including a bake cycle), place the substrate on the proper-sized vacuum chuck using plastic tweezers.
- Apply enough resist to cover the substrate without having it run over the edges.
- Spin the substrate at 3,200 rpm for 20 seconds.
- After spinning, allow the substrate to sit at room temperature for approximately 5 minutes.
- Place the substrate inside an oven at 90°C (194°F) for 20 minutes.

- Allow the substrate to cool to room temperature, then check the surface for foreign material or an uneven surface.
- If a ground plane is required for the back of the substrate, you can either put resist on the backside or cover the surface with Mylar tape.

Process 2: Alumina Substrates

- Bake the metallized substrate at 125°C (257°F) for 30 minutes.
- Place the substrate in a vacuum chuck in the spinner, and cover the surface with photoresist.
- Spin the substrates for 15 to 20 seconds at approximately 4,000 rpm.
- Allow the substrate to air dry for 10 to 15 minutes.
- Bake in an oven at 90°C (194°F) for 25 minutes.
- If a ground plane is needed on the back, you can either spin a resist coating on the reverse side or put Mylar tape over the metallization.

The processes shown above are typical techniques for applying the photoresist. Although placing a coating on a metal-clad piece of material may seem elementary, it is one of the most important steps to be taken on the way to producing reliable microwave circuits.

5.2.3 Artwork Placement and Exposure

We now have cleaned our substrate and applied a thin uniform coating of photoresist. We next need to place the artwork on the substrate and expose it to ultraviolet light so that the required image is on the substrate.

There is much more to artwork placement than simply laying a negative or positive on the material to be exposed. You will recall that all microwave circuits are based on specific lengths of lines and, more important at this point, specific widths. Consider, for example, exposing a circuit board with a piece of artwork that is very loosely held to the board. Figure 5.4(a) shows this condition. You will not have good line definition because it is possible for light to get under the artwork, which will cause the width to be something other than what it was intended to be. This is similar to applying masking tape to an area that you are painting. You need a straight line for definition between two areas. The logical thing to do is put a piece of tape down to separate the areas, but if the tape is not put down firmly, some paint seeps under the tape to form a ragged edge.

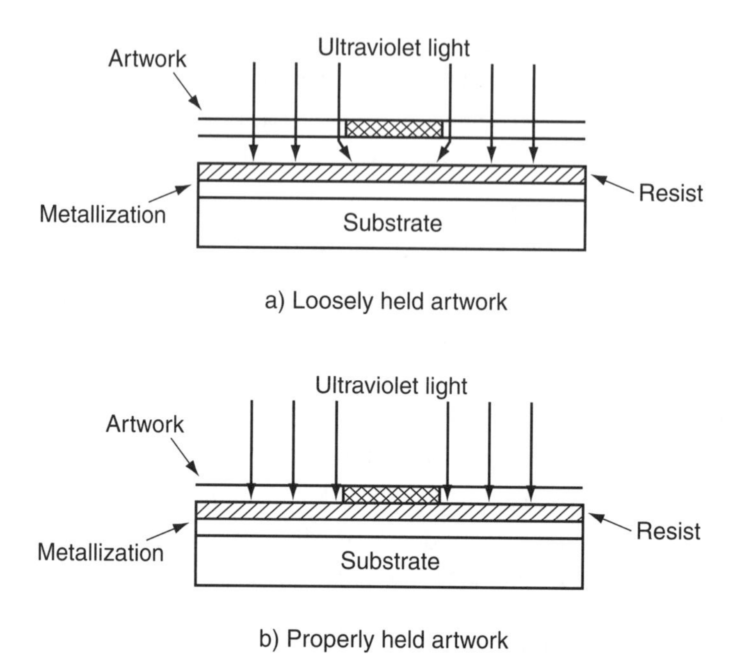

Figure 5.4 Artwork placement.

When the artwork is pressed firmly to the material, as in Figure 5.4(b), the light cannot get under the artwork. It exposes only what was intended to be exposed, producing good line definition that allows the circuit to perform as expected. Thus, a vital step in placement and exposure of artwork for microwave circuits is a vacuum that holds the artwork firmly in place.

The artwork (or mask) to which we have been referring may be made of Mylar or glass, as previously stated. Whichever type is used, the same importance is placed on having the artwork make firm, tight contact with the material being exposed.

Generally, the procedure for aligning and exposing artwork (mask) is much the same from system to system. You should consult the manuals for your particular system for the proper settings to be used. Once you have these settings, the general procedure listed here can be used.

- Make initial settings on the alignment system as specified by the appropriate manuals (power, timer setting, etc.).
- Place artwork (mask) on the proper fixture with the emulsion side down. Do not touch the artwork with your fingers. This will leave fingerprints that may cause exposure problems. Use nonmetallic

tweezers or finger cots. The emulsion side can be distinguished from the nonemulsion side because the emulsion side is dull and can be scratched. You will also notice that the circuit is raised somewhat on the emulsion when viewed through a microscope. (The fixture you use may be one supplied with the system or one fabricated specially for your application.) The emulsion is placed down so that you have the circuit impression in very close contact with the material that you are attempting to expose. This ensures that if proper procedures are used, you will have the proper width lines exposed on the substrate. The film thickness with the emulsion side up could cause a variation in line width similar to that experienced by not having the artwork placed tightly against the substrate, as discussed previously.

- Place the fixture, with artwork, in the alignment system and apply the *minimum* vacuum. This is vitally important if you are using a glass mask because too high a vacuum will shatter an expensive mask and upset a number of people. Begin with just enough vacuum to hold the fixture in place.
- Check for alignment of the mask on the substrate by viewing through a scope. Move to a proper alignment with either minimum or zero vacuum, if necessary. Alignment should be made with either corner or edge marks on the mask or artwork. (Figure 5.5 shows both edge marks and corner markings on a typical mask.)
- When alignment is completed, apply the appropriate vacuum. (Consult the system manual for the proper vacuum for each type of mask.) Push the table with the fixture, mask (artwork), and material into the machine. Previous settings will determine the exposure time.

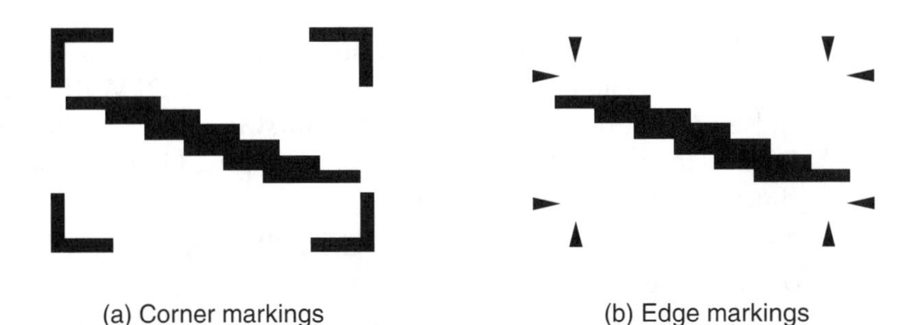

Figure 5.5 Alignment markings.

- Following exposure, the table will either eject automatically or need to be manually removed. When removed, turn off the vacuum and take the mask off of the substrate. The material is now ready to be developed.

The developing process is designed to wash away the unexposed photoresist. There are two basic methods for developing microwave substrates: (1) immersion and (2) spin spray. The immersion is exactly what the name implies: developing and rinsing the substrate by immersing it in the proper solvents. A typical process is outlined here:

- Place the exposed substrate into a dipper, using nonmetallic tweezers, with the exposed side up.
- Immerse the substrate in the developer, and agitate for the time specified by the manufacturer.
- At the end of the required time, remove the dipper from the developer, and remove the excess developer by means of a blotting process.
- Immerse the substrate in the appropriate rinse, and agitate as appropriate.
- Remove from the rinse, and blot any excess rinse from the substrate.
- Blow dry the substrate using dry nitrogen.
- Place the substrate in a 100°C oven for approximately 10 minutes.
- Following baking, the substrate should be checked for foreign materials, voids in the pattern, and the appropriate line width.

The spin-spray process is now outlined using the Solitex spin-spray machine:

- The initial procedure uses a test substrate with no solutions in the machine to set up rpm rate and time sequences. The DEV, RINSE, and SPIN buttons are depressed; the run switch is placed in the DRY RUN position; and the test substrate is placed in the vacuum chuck and the vacuum turned on.
- The START button is pushed and the DEV rpm is adjusted according to specification.
- When the machine cycles to the rinse mode, adjust the RINSE knob for the appropriate rpm.

- When the machine cycles to the spin dry mode, adjust the SPIN DRY knob for the appropriate rpm.
- The spin times are then set for DEV, RINSE, and DRY.
- The switch is now put in the RUN position.
- The developer container is filled with the appropriate developer.
- The rinse container is filled with the appropriate rinse.
- The system is now ready to develop the exposed substrate.
- Place the exposed substrate on the spinner chuck using nonmetallic tweezers or finger cots. Turn on the vacuum.
- The START switch is now pushed, and the system will automatically go through the sequence of DEV, RINSE, and DRY previously set with the test substrate.
- When the cycle is complete, release the vacuum and remove the substrate.
- Place the substrate in an oven at 120°C for PTFE/glass laminates and 150°C for metallized alumina substrates.

With either the immersion or spin-spray technique completed, the substrate or laminate is now ready for the step that will result in the printed microwave circuit: *etching.*

5.2.4 Etching

The etching process removes the unwanted metal from a substrate or laminate, resulting in the originally designed circuit. Metal can be removed in either of the two ways previously used for resist development: immersion or spraying.

Before proceeding any further with discussions on etching microwave circuit boards, we should stop and discuss safety procedures that should always be in place before beginning any etching process. First, the operator should always wear eye protection. These should be chemical-type goggles. Second, general-purpose rubber or neoprene gloves should be worn. And, finally, a rubber or plastic apron should be worn over normal work clothes. These precautions will prevent serious injury during the etching process. Remember that you are working with a solution designed to disintegrate copper or other metal from a circuit board. Imagine what that solution would do to your clothes, your hands, or, more importantly, your eyes. A few seconds of preparation can prevent serious injury.

To illustrate the basic methods for etching a microwave circuit board, we will present generic etching procedures. It is up to the actual operators to tailor these processes to their actual application.

The first process is based on a ferric chloride etching procedure. This was mentioned at the beginning of the chapter. One thing to stress before you do any etching with a solution of ferric chloride or similar chemicals: Do *not* use any metal containers to etch a circuit. In particular, do *not* use anything aluminum when dealing with ferric chloride. If there is any question as to why you should not use aluminum on ferric chloride, a simple experiment will convince you. Pour a small amount of ferric chloride in a glass or plastic tray, take a block of aluminum that is at least 1 foot long, and place one end in the liquid ferric chloride. Within a short period, you will see a violent chemical reaction that will basically "melt" the aluminum and produce a great amount of heat due to a reaction between aluminum and ferric chloride. The aluminum piece will also become very warm when this reaction begins. After doing this experiment, you will no doubt agree with the statements made previously about not using aluminum and ferric chloride together.

With the warning presented and substantiated, we can proceed with a basic procedure for using ferric chloride or, in general, liquid etchants to etch a microwave circuit. Ferric chloride is a good choice for etching circuit boards because it has an indefinite shelf life and can be used repeatedly until it has absorbed as much material as it can hold. This is an important point. Most people who do not get involved with the etching of printed circuit boards do not realize that when you etch a circuit board, the copper must go somewhere. It does not simply evaporate into thin air. It goes in the etchant. There is a definite capacity for each etchant and when that capacity is exceeded, the etching properties of the solution are significantly diminished. Thus, this is a parameter of your etchant that should be understood.

The equipment needed for such an etching procedure is a hot water source and sink, plastic or glass trays if the etching process is to use a tank or tray, and a suitable spray etcher if the technique is to use spray etching.

The time required to etch a particular circuit depends on different factors. These factors are:

- *Agitation.* Agitation of the liquid ensures that there is fresh etchant to the board at all times. The more the etchant is agitated and moved around, the more efficient the etching process will be and the less time it will take to etch the circuit board.

- *Heat.* An etchant will react significantly faster when temperatures are elevated to 105°F (40°C). To prevent any excess fuming on the board, do not exceed 115°F (46°C). (Fuming basically means to emit fumes from the etching circuit board.)
- *Copper.* There are basically two areas that involve copper. One concerns the absorption of copper as the etching process takes place. As more copper is absorbed by the etchant, the rate of etching is slowed with a longer etching time resulting. The second is the total amount of copper that the etchant is required to etch from the circuit. If, for example, you have a circuit board being etched that had 2 oz copper on it, it would take much longer than a piece with only ½ oz copper on it. The smaller quantity of copper would etch much faster. As a matter of fact, narrow lines and gaps on microwave circuits are usually constructed and etched with very thin copper so that the etching process is swift and efficient. For most applications, it is recommended that the etching process be as quick as possible to prevent circuit undercutting and removal of the entire transmission line due to overetching.

With some of the basic facts presented, we will offer a short procedure for etching with ferric chloride, first in a tank or tray and then in a spray etcher.

1. Pour the etchant into the tray or tank to cover the circuit board by at least ½ inch.
2. Raise the temperature to a value between 105°F and 115°F. This can be accomplished by either placing the tray in a hot water bath or using infrared lamps over the tray. Remember not to exceed the 115°F value to be sure that fuming does not occur.
3. Move the circuit board around within the etchant to produce an agitation of the etchant. This may be accomplished by actually moving the tray around or by injecting air into the solution with a small pump. This agitation will also decrease the etching time for the circuit.
4. Following the etching of the board, remove the board from the solution and thoroughly rinse it with water. If the etchant leaves stains on the board, you may need to brush the surface with a soft bristle brush to remove the etchant stains from the circuit board. Removal of the etchant from the circuit is particularly important

when the circuit board is to undergo soldering or plating procedures later on. Any etchant left on will impair the process and make the following operations very difficult.

5. After the circuit board has been thoroughly cleaned, the etchant itself must be taken care of. You should pour the etching solution into a plastic container and save it for the next use, being sure that the solution you are pouring back is still capable of absorbing copper for future etching operations. Do not mix fresh and used etchants. Although they will do the job for you, it is not a good idea to mix the two because it will cut down on the overall efficiency of both the fresh and used etchants and the etching times will increase greatly.

Spray Etcher

When you use a spray etcher, you basically only need to place the ferric chloride in the etcher, place the circuit board in the proper position, *consult the manufacturer's manual for specific etching times*, and turn the machine on. Following the etching process, you should, once again, rinse and clean the circuit board. The etchant in a spray etcher is changed periodically, depending on how much etching was done with the present amount of etchant in the machine. Proper disposal of the chemicals when the etchant is changed is very important. You cannot just pour them down the sink and hope nobody sees you do it. Contact the Environmental Protection Agency and find out where there are approved hazardous waste disposal areas or companies. All chemicals will have the hazardous content included with the paperwork that comes with them.

Whenever dry chemical etchants are used to etch printed circuit boards, there is always a mixing process required to turn them into a wet etching material. The mixing processes are determined by the manufacturer and should be followed so that the maximum efficiency can be obtained during the etching process.

Dry chemical etchants have distinct advantages over etchants such as ferric chloride. They will not stain clothing, tanks, or your skin if the solution gets on your hands; dry etches rinse more easily and leave no residue in plain water; and they are faster and maintain a relatively even etching rate throughout their entire mix life. As with any product, there are also disadvantages to dry etch chemicals. They have a shorter active life than wet chemicals and will more readily attack natural fibers such as cotton, wool, and linen. The type of chemical etch that you use depends on your application and the equipment available to you to actually do the etching.

Before continuing further, it might be a good idea to look at the following glossary of terms commonly used when talking about etching:

Glossary

Aqueous-based: Chemicals shipped in a dry state that need water added to them prior to use.

Exposure: Subjecting photoresist to ultraviolet light.

Negative: Artwork or film in which the desired pattern is clear on an opaque background.

PCB: Printed circuit board.

Photoresist: A light-sensitive coating that is applied to a circuit board prior to etching.

Photosensitive: When a material's properties are altered when exposed to light.

Positive: Artwork or film where the desired pattern is opaque and the background is clear.

Print-Etch: The process where the desired pattern is applied (print) to a circuit board material and the unwanted copper is removed (etch).

Resist: A material applied to the circuit board material to protect the copper from the etching solution.

UV: Ultraviolet light.

One last step needs to be performed before checking all of the dimensions on an etched substrate and before checking for cracks in the lines or holes in the traces. This step is the removal (or stripping) of the resist that we took such great pains to put on the substrate earlier. This is handled as follows:

- Under a hood for ventilation of the area, heat the appropriate resist stripper to 100°C.
- Immerse the substrate in a holder into the stripper with the pattern side up.
- After 2 minutes, check the pattern for resist by rinsing it with a spray of hot water to remove the suds.

- Any remaining resist may be removed by using a cotton swab dipped in stripper.
- Rinse the substrate in a hot water spray again, and place it in a rinse tank with running deionized water for approximately 10 minutes.
- Blow dry with nitrogen.

We now have our completed substrate, which we will check for any defects. Problem areas that should be inspected are:

- Cracks in the lines;
- Pits in the metallization;
- Rounded corners on the lines instead of good mitered corners;
- Proper etching (good sharp edges on copper boards and only gold showing on multimetallized substrates);
- Proper width of lines (minimum amount of undercutting).

You will note that the last statement says a *minimum* of undercutting. There is always a certain amount of undercutting of the metallization because it is not possible to have an etching process stop on demand. There is always a certain amount of creeping of fluid that will etch away some of the line beyond where you may want it to be. This undercutting is shown in Figure 5.6. You can see in Figure 5.6(a) how sharp the edges are on the exposed area of the substrate. Figure 5.6(b), however, reveals the real-world finished product, which is undercut. A good rule to use when gauging etching procedures is that 1 mil undercut is acceptable. Anything larger than that would have to be considered with regard to your particular application. If line widths are critical, anything over 1 mil (0.001″) would have to be rejected. If your lines can tolerate 1.5–2 mils of undercut, you can pass many more substrates through inspection.

Another type of etching that is finding more applications is the *plasma etching* technique. This method uses gasses to perform the etching process. As an example, a multilayer application has its gas mixture explained as follows:

- The gas used for most plasma desmear and surface treatments is a mixture of 70% to 90% oxygen with the balance being tetrafluoromethane and nitrogen. This mixture provides a rapid desmear and etch-back. For surface treatment of through-holes for plating, a mixture with less CF_4 is desirable due to fluorine residue remaining on the surface.

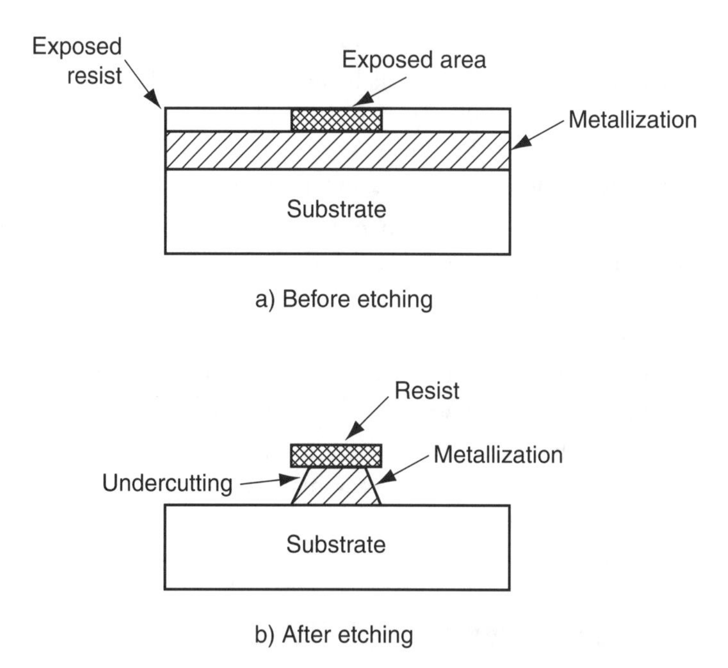

Figure 5.6 Undercutting.

- Oxygen has some etching capability, but pure oxygen plasmas have not been found suitable for desmear due to their very slow etching rate and because they quickly form films on copper.
- Another etching mixture in use is nitrogen trifluoride and oxygen. This mixture has three times the etching rate of CF_4O_2 and has been used successfully for surface treatment of PTFE panels prior to copper deposition. We should mention that NF_3 has not been totally accepted because of its toxicity.

You can see that the method of plasma etching is a viable one to use for microwave circuitry. The example given above is for multilayer circuitry, but a single layer may be etched in much the same manner. This process is often used when a chemical etch, as has been previously explained, would be harmful to the materials being etched.

In conclusion, we can give some general guidelines for etching of microwave laminates:

- A gap in a circuit is more difficult to image and to etch than lines, so specify the gaps to be larger than the associated line widths.

- Features smaller than two times the resist thickness are difficult to image. With standard resists being approximately 0.002″ thick, images of less than 0.004″ may be difficult to achieve. You are well advised to find out the thickness of the resist being used so that the minimum line widths can be determined.
- Etching of lines smaller than two times the copper weight is also difficult. If ½ oz copper is used, the minimum line width would be 0.0014″. Similarly, for 1 oz copper, the minimum line width is 0.0028″.
- Sharp corners where changes in direction occur are more difficult to image and to etch than gradual curves or 45° mitred corners.
- Be aware that there are a variety of methods of specifying the measurements of line widths. They may be flair to top surface, flair to flair, or top to top.

5.3 Plating and Plated-Through-Holes

If you look up the term *plating* in a dictionary, you will find that it says: "A coating of gold, silver, or other material." That is exactly what we are referring to when we mention the plating of a microwave circuit board or the plating-through of a drilled hole on a circuit board. It is the coating of that surface or hole to either protect the copper traces that were etched or to provide a continuous path between the top and bottom of the board. We will look at both the plating and plated-through-hole procedures in the next two sections.

5.3.1 Plating

As mentioned above, plating is the coating of a material with another material. In the cases we will discuss, we will coat a microwave circuit trace with a metallic covering designed to protect the copper traces from oxidation. If you have ever placed an etched circuit or a piece of copper-clad microwave material on a table for any period of time, you know that the nice shiny piece of material quickly looks dingy and unappealing. The material's appearance is actually a secondary concern when it comes to a microwave circuit. The real concern is the oxidation process that increases the resistance of that copper that was so pretty a short time ago and thus produces a higher copper loss for that circuit.

To eliminate the oxidation process, we simply cover the copper with a material that will not oxidize and will have low loss characteristics. Common materials are gold, silver, or tin-lead, which are placed over the copper traces. All of these materials exhibit low loss characteristics and can be plated over the copper traces with relative ease.

One point that should be made before proceeding further with the discussion on plating is that of the thickness of the plating that is finally put on the circuit boards. This layer of metal should be very thin. This is an absolute necessity since the microstrip and stripline circuits rely very heavily on what the thickness of the copper, or total trace, will be. When calculating the width of a transmission line for a certain impedance, one of the parameters to be used is t/b. That is, the thickness of the conductor (t), divided by the ground plane spacing (b), or the thickness of the material. If calculations are made with 1 oz copper, for example (which is 0.0014″ thick), the impedance of the transmission line would not be as calculated if you placed a 0.002″ thickness of plating on top of that 1 oz copper transmission line. You can see, therefore, how important it is to keep the thickness of any plating to an absolute minimum. Many times, the thickness is in the micron range (10^{-6} inches). This sounds very small, but realize that the only thing this is designed to do is cover up the copper traces to protect them from the environment. Being very thin also ensures that the loss due to the plating is kept at an absolute minimum.

As in any circuit board operation or procedure, the cleaning of the material prior to performing the operation is very important. If we relate back to the cleaning discussions previously presented, you should realize just how important it is to have a very clean surface to plate to. Any contaminate that is present will not be a good surface for plating.

To emphasize how important it is to properly and thoroughly clean material before plating operations begin, we will repeat some of the cleaning information presented in Section 5.2.1.

To be certain that you are using the proper chemicals to clean a particular material, you should consult the manufacturer's recommendations. Knowing the *proper* cleaning method for your particular application, like which commercially available chemical copper cleaners to either wipe on the board or dip it in and then rinse and dry it, will greatly enhance your chances of producing an accurate and usable circuit board. We will not recommend specific methods of cleaning because there are variations that could cause you more problems than we are attempting to solve. Suffice it to say that all cleaning processes consist of three basic operations: (1) solvent bath, (2) rinse, and (3) baking (or drying). The number of steps may vary; more than one bath or one rinse may be involved, as is the case when hard substrates (alumina, for example) are prepared for metallization by a sputtering process. This requires much more than the basic cleaning process previously mentioned.

To understand the steps of a cleaning process, it helps to know what is being cleaned in the first place. There are two basic types of contaminates that must be removed in order to achieve a good etch: oil-based soils and water-soluble soils. Oil-based soils can be removed by vapor degreasing or using an ultrasonic cleaning process that basically shakes the contaminates off of the material. Water-soluble soils are removed by using deionized water or high-purity alcohol.

The deionized water, or DI H_2O, is a water that has been purified by removal of ionizable materials. Ionized materials are those that have electrons easily removed from molecules. This removal causes ions to be formed and an increase in free electrons that will reduce the electrical resistance of the material. For a thorough cleaning process, these materials should be removed from the cleaning solutions. Thus, deionized water is a primary cleaning product that should be available.

In order to fully understand the cleaning of a circuit board before plating it and producing a usable finished board, one point should be presented and emphasized greatly: do *not* touch the substrate or laminate with your hands. Use nonmetallic tweezers or rubber finger cots. Your fingers can leave a film of oil or dirt particles on the material that defeats everything you are trying to accomplish with the cleaning process. It can be rather discouraging to clean your circuit board, pick it up with your hands to move it to another location, go through the plating process, and then find out that there are discontinuities in your plated circuit that look exactly like the fingerprint of your right thumb. When handling laminates or substrates, use the proper precautions and save time and trouble.

The process of cleaning circuit boards is one that should not be taken lightly. It should be apparent at this point just how important a good cleaning is to prepare the circuits for etching. To review the cleaning process, we will summarize the Do's and Don'ts as they pertain to the pre-plating process—that is, how to clean a metallized substrate or copper-clad laminate prior to plating.

1. *Do not* touch a substrate or copper-clad laminate with your fingers. Use a set of nonmetal tweezers or finger cots.
2. *Do* use a liquid cleaning solvent. Avoid abrasive cleansers because they may scratch the metallic surface. These fine-grained scratches will cause problems with your microwave circuit.

3. *Do* use vapor degreasing or an ultrasonic cleaning for oil-based soils that may be present on the surface. Use deionized water or high-purity alcohol for water-soluble soils.
4. *Always* end the cleaning process with a baking step that removes any films left from previous cleaning operations.

With the topic of cleaning covered, it is now time to look at the actual plating of microwave circuits.

Typical finishes used in microwaves are tin-lead fused and unfused, tin, gold, nickel, and nickel/gold. All of these plating materials are common because they can be used as an etch resist for ammoniacal etching chemistry.

Probably the most popular and widely used plating material is tin-lead. Conventional tin-lead plating of 200–300 μ in (0.0002″ to 0.0003″) is more than adequate for microwave frequencies. You should be aware, however, that as the frequency of operation increases, the skin effects can take over and may cause adverse effects on the circuit. Recall that we previously spoke of the t/b ratios and said how the plating must be basically invisible electrically to the circuit. When the skin effect comes into play, there may be too much thickness present to allow the circuit to operate properly. When this occurs, the thickness of the plating needs to be reduced. Generally, for most applications, a 300 μ in matte tin plating is the best finish available. It has excellent solderability and a long shelf life after plating for soldering. It also provides excellent protection against environmental effects when put in a system in the field.

Probably the second most popular method of plating to use is gold. Gold is a very good conductor, which is an excellent property for a plating material, and it has excellent corrosion resistance.

Gold plating is a method of plating that finds many applications throughout the microwave field. Even though it is more costly than other methods, it is the choice of microwave designers many times. Although gold is a nice surface to look at and can be found in many circuit applications, individuals who do the plating operations can encounter a variety of problems when using gold.

One of the prominent problems that arises is due to the fact that gold cannot easily be etched in the presence of copper. When the two are etched together, an accelerated etching process takes place on the copper that results in very poor edge definition on the circuit.

One method used in an attempt to reduce some of the gold plating process is an electroless plating process. This does not require an external electric field, and all isolated areas of copper will be uniformly plated with the gold. Since the circuit is first etched in bare copper, the best possible resolution can be achieved. This sounds like an excellent solution to the gold plating process. There are, however, some drawbacks to this process for certain applications. The gold deposit for such a process is very thin (20–30 μ in) compared to the standard tin-lead plating discussed earlier. It also has a much lower purity than other gold plating processes (around 97%). Since the gold layer is so thin, the corrosive protection is not as good as that of other methods that use a thicker covering.

Probably the most general method for gold plating and the subsequent etching of the circuit is called pattern plating. In this method, a negative image of the circuit is formed on the surface of the copper using the proper photoresist. Gold is then electroplated across the exposed area to the needed thickness. The next step is to strip the photoresist from the material and etch the copper from the board to reveal the circuit. One problem that occurs in this process is the undercutting of the edges. This causes problems with the circuit and, in particular, the impedance of the transmission lines that have been etched. There are methods being worked on that will apply chemistry to alleviate this problem. But at this point, you should compensate for any variations used so that you get as close as possible to the final etched width, even when there is gold plating on the circuits.

To sum up plating, look at Table 5.1.

Table 5.1

Metal	Resistivity ($\mu\Omega$ cm)
Silver	1.62
Copper	1.72
ED Copper	2–3
Gold	2.45
Aluminum	2.80
Zinc	5.80
Brass	7–8
Nickel	7.80
Platinum	10.00
Tin	11.50
Lead	22.00

5.3.2 Plated-Through-Holes

As previously stated, the idea of plating through a hole on a Teflon material was unheard of some years ago. The data sheets for Teflon material made a special point of emphasizing that so no one would attempt it. Today, plating-through-holes on Teflon-based materials and every other kind of material is accomplished every day and is part of microwave life.

Why do we need to plate-through-holes on microwave circuit boards? Generally, we are trying to get a ground connection to the top of the board, or attempting to get from one layer to another. An excellent example of the first criteria is shown in Figures 5.1 and 5.2, which we previously discussed. This circuit is an interdigital filter circuit. Each of the resonators in this filter is required to have a ground connection to them. The problem arises due to the fact that the ground connections are on opposite ends of the transmission line resonators. The first is at the end of the input line, the next is at the other end, and it continues to alternate. If the grounds were all at the same end of the circuit, a common ground could be placed on the circuit. But since there is an alternating pattern of grounds, this is not feasible. The solution is a plated-through-hole at the end of each resonator connected to the ground plane on the other side of the board. This is an excellent solution to a problem that kept the interdigital filter from being used for many applications.

We will go through a few steps for plating-through-holes (edge plating of a circuit board falls into this category) and point out some areas where care should be taken when performing this process.

As stated before, the plating of Teflon-based material was never done but is now a common process performed every day. When Teflon fiberglass materials are plated, there may be some fibers present in the hole after it is drilled. These fiber bundles produce an uneven surface in the hole and make it difficult to plate to. A hydrofluoric acid etch before plating will get rid of these fiber bundles and produce a very smooth edge to plate to. If no fiberglass fibers form bundles in your material, this step is not necessary.

If you are going to plate holes or edges with copper, you must take certain precautions to either avoid or reduce smear on the machined surfaces. Smear is just what the name implies, a sloppy hole in which to attempt to plate through. The problem with smear is that it will interfere with the electrical continuity between the copper in the inner layer of a multilayer board and the copper of those multilayers. If you only have a two-sided board, the problem is limited to interference with a good copper bond to the side of the hole. Smear can also be removed by a process called vapor honing. This involves directing abrasive particles carried in water or air streams to be forced against the surface, which will remove the largest percentage of smear and create a much smoother area for plating.

When plating holes or board edges, a layer of electroless copper is required. This may be the required thickness or just thick enough to provide a conductive base for electroplating more copper to achieve the required final thickness. When depositing the electroless copper from a copper-formaldehyde solution with a carefully controlled stability, it is usually initiated on a surface by a layer of colloidal palladium particles that were previously deposited from a commercially available aqueous solution. If you intend to have the electroless copper stick to the material, the surface must be water-wettable. This is commonly attained by treating the circuit board with an organic solvent solution of metallic sodium, which is available from a variety of manufacturers.

In order to produce an efficient plated-through-hole, it is necessary to reduce, or eliminate, any voids or open spots in the hole wall. Some factors that cause such voids are: air bubbles trapped in the hole during certain processing steps; improper cleaning methods at the start of the process; long immersion times during the process; the wrong accelerator time or concentration (the accelerator serves to remove the tin from the palladium deposit); and damage to the hole during drilling.

An alternative to the conventional plated-through-hole process is used for "blind" holes. A blind hole is one that is not drilled completely through. This occurs when you purchase a microwave material with a metal backing on it. This is an excellent ground plane, but how do you get to that fabulous ground? The answer is a blind hole that is drilled through the board and into the metal backing. This hole can then be plated for a good ground connection from top to bottom. The process used for such an application is to sputter the metal into the hole and onto the metal backing. This is exactly like the process covered previously when we discussed ceramic substrates. It is an approach that does the job very nicely.

5.4 Alternate Methods of Producing Circuit Boards

Thus far, we have talked about etching a circuit board that is a true representation of our microwave circuit. This process can be lengthy and must be duplicated every time there is a circuit change. This long time frame is one that many times cannot be tolerated during a design's development stage. To decrease this time and produce a circuit board that is ideal for breadboard work, circuit board plotters have been developed. These devices will take the artwork that we developed in Chapter 4 and produce a finished prototype circuit board in a short period of time. These machines are capable of producing lines as small as 4 mils with 4 mil spacings between transmission lines. They are compatible with Gerber, HPGL, and DXF formats so that the programs discussed previously will operate well with them.

The only drawbacks of these machines are that they sometimes cut the circuits too deep and they are noisy. The problem of deep cutting can be taken care of by properly adjusting the machine. The noise problem is usually caused by a vacuum system designed to keep the air clean of harmful particles that may be floating around. If the machine is put in a room by itself, the noise will not bother anyone but the operator. Some operators start the machines and then leave the room until everything is finished. That's not a bad idea.

5.5 Chapter Summary

This chapter addressed two vital areas for the microwave circuit—etching of the circuit and plating of that same circuit. The etching is very important because the transmission lines for microwave circuits are critical and have to have their widths very closely controlled. It is not good to have a quarter-wave transmission line that varies from 0.025″ to 0.050″ over a short span of that line. Protecting the final etched circuit is also one of the primary functions of the plating process, or the process of plating-through-holes to connect to a ground or different layer. All of these processes are important in the final analysis that makes the microwave circuit perform as originally designed.

6

Bonding Techniques

6.1 Introduction

Now that we have designed a microwave circuit, have artwork to make it a reality, and have etched and/or plated it, we may need to attach components to it and, definitely, attach it to some sort of package. This is where the idea of bonding comes in.

What do we mean when we use the term "bonding"? Basically, bonding is a method used to produce good electrical contacts between metallic parts (although some nonconductive bonding materials are used to attach microwave chips to a base substrate). It requires methods that will join surfaces and produce good electrical contact that may be made in a variety of ways. There is no one bonding method that should be used for every application.

The methods presented in this chapter are solder, epoxy, thermocompression bonding, thermosonic bonding, and ultrasonic bonding. As previously stated, there is no one method that will work for every application. There is, however, one or more methods that will work for your application.

Before proceeding to the individual methods for microwave bonding, we should once again bring up an important topic that cannot be overemphasized. That topic is *cleaning*. It is vitally important that the surfaces to be bonded are as clean as possible prior to the bonding process. We presented numerous examples of trying to perform operations where the surfaces involved needed to be clean. Painting an unclean surface and taping a window with dirty masking tape are but two of them. In each case, we emphasized that all dirt and oil must be removed from the surfaces before attempting to perform any operation. The cleanliness of the surfaces helps determine if the operation lasts only for the first few minutes or for the lifetime of the unit being operated on.

This holds true for bonding two materials together. Whether it be a circuit board with chip resistors soldered to it or a circuit board that must be attached securely to a ground connection in a chassis, the security of the bond is very important. In addition, many times there are environmental considerations to take into account (temperature, humidity, etc.) or there may be mechanical considerations (vibration, for example) that will stress the bond that you must make. These factors should also be taken into consideration when choosing a bonding agent for your circuit.

6.2 Solder

The term *solder,* unfortunately, is one that practically everyone recognizes, and its familiarity allows for predetermined ideas about what solder is and how it is used. So many people think of the soldering process as one in which an iron or torch is used to heat the area to be joined and solder is fed to that area. The general consensus often is that the more solder put on, the better the connection is: as long as the solder keeps melting, keep feeding it to the joint. This usually gives a connection, but its reliability can be debated. The difference between a soldering method and the *right* soldering method is one that should be clarified.

Figure 6.1 shows the requirements for the right solder connection. We will investigate each of these requirements, providing you with the information needed to make the decision about which method is best for your application.

Our discussion begins with a definition of the term "solder." (We will be discussing both tin-lead and indium-based solder as well as other alternate solders.) It is generally defined as a meltable metal or alloy that joins two metal surfaces. The melting point of solder, of course, must be lower than that of the metals it is joining. The solders used for microwave application are termed *soft solders* because of their low melting points: below 450°C (842°F) and usually well below 300°C (572°F). Ordinary tin-lead (SnPb) solder melts at 190°C (375°F), and most indium solders, which are common in microwave circuits, generally melt at 150°C (300°F) or lower. There are also *hard solders* that have melting points above 450°C; these obviously are not used for microwave circuits because they would cause destruction of the components being attached long before the solder even began to melt.

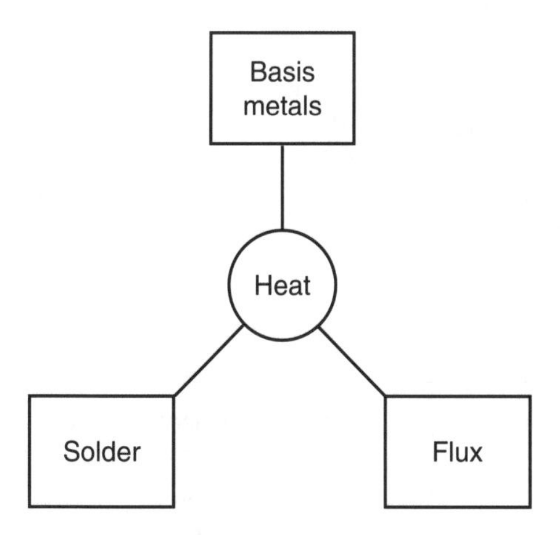

Figure 6.1 Requirements for a good solder connection.

The melting point of a particular solder depends on its composition. Consider common tin-lead solder. Most people take this solder for granted because it is the most common type available and has so many uses. The most widely used tin-lead solder (60/40) has a melting point of 370°F (187.7°C). If you use a tin-lead solder with a 50/50 composition (50% tin and 50% lead), you increase the melting point to 417°F (213.8°C). If you go one step further and use a 40/60 composition of tin-lead solder, the melting point becomes 460°F (237.7°C). A small variation in content of the solder can cause significant changes in the temperature at which it melts. This, of course, is also true of solders with different compositions of indium and other metals in them.

Indium, as discussed in Chapter 3, is a soft metal that has many properties beneficial for microwave applications. One property of indium that makes it useful for solder application is its low melting point. Consider, for example, the melting point of indium as compared to tin and lead. This comparison is shown here:

Element	Melting Point
Indium (In)	156°C (312.8°F)
Tin (Sn)	231.9°C (449.4°F)
Lead (Pb)	327.4°C (621.3°F)

There are two observations that can be made from the listing above: (1) the melting point of indium is significantly less than either tin or lead, and (2) it now is apparent why the melting temperature of tin-lead solder increased as the lead content increased. (Remember that 60/40 melted at 187.7°C and 40/60 melted at 237.7°C, a difference of 50°C, simply because of the increased content of the high-melting-temperature metal: lead.)

To realize how much lower a temperature indium solder will take to melt, consider the direct comparison here:

Solder	Melting Temperature (°C/°F)
Sn/Pb (50/50)	213.8/417
Sn/In (50/50)	117.2/243

This comparison shows that the same 50/50 combination of tin and either lead or indium results in more than 96°C difference in the melting point. The lower, of course, is using indium. This is a significant difference.

We have presented two basic types of solder thus far: tin-lead and indium-based solders. We have concentrated only on melting temperatures to this point, but there obviously are good and bad points other than temperatures associated with each type of solder. These points will be our next topic of discussion.

Tin-lead solder, as previously mentioned, is available in a variety of compositions. Most texts on solder list no less than six combinations of tin and lead solder:

- 63/37
- 70/30
- 60/40 (the most common)
- 50/50
- 40/60
- 20/80

(*Note:* The numbers indicate the percentage of each metal, tin and lead.)

Melting temperatures for these compositions are as follows:

- 63/37—361°F (182°C)
- 70/30—367°F (186°C)
- 60/40—370°F (187°C)

- 50/50—417°F (214°C)
- 40/60—460°F (238°C)
- 20/80—531°F (277°C)

The type of tin-lead solder that you use will depend on your application as far as such things as metallurgical makeup of the devices to be connected and the temperature limits of these devices. For example, to use 40/60 tin-lead solder on a transistor that had a maximum temperature range of 200°C would not be wise because it takes a 238°C temperature to melt the solder. The temperature range of the devices you are soldering must always be considered.

Similarly, as mentioned previously, you should be aware of the metallurgical makeup of the devices that you are soldering. Tin-lead solder should not be used to solder to gold or silver plating because the tin tends to scavenge gold and silver. You will recall that gold could not be put directly on the alumina substrates because of a scavenging process; an adhesive layer was needed to make the complete substrate. This same "leaching" process will occur with tin-lead solder on gold or silver plating unless the solder has small amounts of gold or silver to reduce this phenomenon (62Sn/36Pb/2Ag, for example). A very brittle AuSn (gold-tin) intermetallic layer will form, which usually causes connection failure in extended temperature cycling. Therefore, pure tin-lead solder should be used only when the contacts on a component or the ground plane of a substrate has a tin plating or is of metallically compatible construction. Even though we said that tin-lead-silver (62Sn/36Pb/2Ag) solder can be used on gold, it is a good idea to stay away from that process and use an indium-based solder instead, just to be safe.

Tin-lead solders are higher-temperature "soft" solders that find uses in microwaves to attach components (chip resistors and capacitors) and also to attach substrates to cases. When using tin-lead solder, however, remember to check the metals that are being soldered so as to prevent a brittle connection that could cause problems later.

We have covered in detail the lead-based solders used for many microwave applications. Before proceeding to the indium-based solders, we should point out that lead-based solders are being replaced in some areas because of environmental factors. The elimination of lead is an area that some people and companies are striving to accomplish.

If the lead-based solders are replaced, what will they be replaced with? Some of the solders that have been proposed are: bismuth-tin-iron (BiSnFe), tin-silver-zinc (SnAgZn), and tin-antimony (SnSb). Interestingly enough, all of the proposed substitutions have shortcomings as compared to tin-lead. These shortcomings are price, physical properties, metallurgical properties, or mechanical properties. It is also interesting to note that the most promising replacement for lead-based solders is one that we will look at in our next section, tin-indium (SnIn).

Indium solders are generally lower-temperature solders that are much more flexible than tin-lead. This flexibility is due to the softness of indium (described in Chapter 3), which generally is an advantage but can cause problems by being too soft.

Typical solders with an indium base are tin-indium (50Sn/50In), lead-indium-silver (15Pb/80In/5Ag), and lead-indium (50Pb/50In). Although there are many more types of indium solder, these are the most common types used for microwave applications. Note that Indium Corporation of America in Utica, New York, offers a variety of kits containing different types of indium solder. There is one offered that is called the *microcircuits kit.*

One thing to notice about most of the Indium solders listed is that the only elements used are indium (In), lead (Pb), and silver (Ag). There is an obvious lack of tin in any of the solders because, as you recall, the tin reacts with gold to form AuSn metallics that are very brittle and cause connection failure. Because most microwave components use a gold plating for cases and conductor coating, you do not want tin in the area to produce this condition.

As an example of actual applications of indium solder, you could use the 50In/50Pb to put substrate onto a carrier plate or case bottom and use an 80In/15Pb/5Ag to attach the components to the substrate. This would put less thermal stress on the component (temperature of 149°C) and use the higher-temperature solder (temperature of 209°C) for the larger ground-plane area. In addition, the higher-temperature solder would be used so that the ground plane would not come unsoldered when heat was applied to attach the components.

Before leaving the topic of indium solders, we should mention that just as tin and gold should not come in contact with each other, indium and copper should also be kept apart. If contact is made, the copper will diffuse into the indium and cause an unreliable connection. This interface should therefore be avoided.

So, we have looked at the first block of Figure 6.1: solder. The previous discussions covered two of the most prominent types of solders used in microwaves: tin-lead and indium-based solders. Table 6.1 is a summation of some of the typical solders used for microwaves. They are listed by composition and in increasing order of melting temperature.

With the topic of solders completed, the next block from Figure 6.1 is that of *flux.* The main job of solder flux is to remove impurities from the metallic surface to be soldered so that the soldering process can take place as

Table 6.1
Microwave Solders

Composition (%)	Melting Point (°C/°F)
50 Sn, 50 In	125/257
15 Pb, 80 In, 5 Ag	149/300
100 In	157/315
63 Sn, 37 Pb	182/361
70 Sn, 30 Pb	186/367
60 Sn, 40 Pb	187/370
50 Pb, 50 In	209/408
50 Sn, 50 Pb	214/417
40 Sn, 60 Pb	238/460
75 Pb, 25 In	264/508

Sn = Tin
Pb = Lead
In = Indium
Ag = Silver

efficiently as possible. The impurities that it removes are usually oxides that have formed on the metal surface. A common oxide is one that is familiar to all of us: iron oxide, or rust. The flux ensures that oxides or other impurities do not interfere with the soldering process. To be an effective "cleaning" agent, a flux should accomplish the following tasks:

- Provide a liquid cover over the material to be soldered and shut out any air up to the solder melting temperature. (Any air will increase the creation of oxides on the metal.)
- Dissolve oxides on the metal surface and carry unwanted material away (basically, make the solder area free of any impurities that will hinder connection of the metals to be joined). This is part of what is called a *wetting* process: the process that forms a uniform, smooth, and adherent film to allow soldering to a base material.
- Be easily displaced from the metal when the solder reaches its fluid state. (Once the area is cleaned by the flux, the flux must move out of the way to allow the soldering process to take place.)
- Be easily removed after the soldering process is complete. (This provides a clean connection.)

There are three types of fluxes used in soldering: corrosive (inorganic), intermediate (organic), and noncorrosive (rosin):

- *Corrosive (inorganic) flux,* designated IA, is not used for electronic assemblies. It is used for metal alloys or stainless steels that are difficult to solder. This type of flux is very corrosive and can damage other components that are in the area you are soldering if great care is not taken. You will not see this type of flux used in microwave or any other electronic applications.
- *Intermediate (organic) flux,* designated OA, is divided into three subcategories: organic acids, organic halogens, and amines and amides. All three subgroups are corrosive (the organic halogens are more corrosive than the rest) and are very temperature-sensitive. Most produce condensed fumes; one must carefully dispose of such fumes to ensure operator safety. These fluxes, like the corrosive fluxes, are not used for microwave or electronic applications.
- *Noncorrosive (rosin) flux* is the most frequently used flux for microwave and electronic application. The rosin flux is divided into three subgroups also: water-white rosin, designated R; mildly activated rosin, designated RMA; and activated rosin, designated RA.

The safest and mildest flux is the water-white rosin (R) type. It is used on very clean surfaces where gold and silver (highly solderable metals) are utilized. Any residue left after soldering creates no corrosion problem. It can, however, be removed with a solvent such as trichloroethane (1,1,1 trichloroethane is the minimum concentration of solvent used; higher-purity solvents are also available for use).

The mildly activated rosin flux (RMA) has a shorter wetting time than R fluxes and is used, once again, on gold, silver, and also copper. These fluxes are noncorrosive and nonconductive and are removed from the finished circuit only in very critical applications such as aerospace equipment. When it is necessary to remove the flux, a combination of alcohol mixed with 1,1,1 trichloroethane can be used.

Activated rosin fluxes (RA) are used on metals such as nickel and cadmium. They are the strongest and most active of the rosin fluxes and are sometimes used to speed up the soldering time of such metals as gold. Residue from these fluxes should always be removed with a solvent such as isopropanol (alcohol) and 1,1,1 trichloroethane.

As should be clear at this point, the choice of flux for a solder connection is at least as important as the choice of solder. In many cases, it is more important, as the wrong flux will not remove the oxides or impurities, may not flow out of the way of the solder, and may leave a residue that will be harmful to either the circuit being soldered or an adjoining circuit. Great care must be taken when choosing the flux to use.

The final block shown in Figure 6.1 is called *basis metals.* These, very simply, are the metals being joined. We previously mentioned metallic combinations that will result in an improper solder connection. Tin and gold or indium and copper are examples of incompatible metals. You should be sure of the metallic makeup of the items to be joined prior to attempting to join them. Manufacturers of chip components (resistors and capacitors) often will list recommended methods of attachment and will give solder types that you should use. Take note of these recommendations, and follow them as closely as possible.

The topic of *heat* in the soldering process has been covered over and over again. You should make this one of your primary considerations when contemplating a solder connection. Some tables have been presented in this chapter, and solder vendors, of both tin-lead and indium, will gladly provide literature with additional tables of temperatures.

We have covered all aspects of the requirements of a good solder joint: solder, flux, basis metals, and heat. Let us now see how these work in an actual application. The best application is one where some problems can occur if care is not taken in choosing the right solders, flux, metal, or heat. That application is *soldering to gold.*

As previously mentioned, soldering to gold is an operation that requires care. If there is a tin content to the solder, a brittle connection results. If the gold layer is so thin that it is removed by the soldering process and there is copper below it when you are using an indium solder, there will also be problems. This can be a difficult connection to make, but if careful consideration is given to all aspects of the connection to be made, it can be a fairly routine matter.

Let us use a specific example. Suppose we have to solder a substrate consisting of chrome-copper-gold metallization to a gold-plated kovar carrier plate. The metallization has the following dimensions:

Chrome—200 to 300 angstroms (Å);
Copper—250 $\pm$ 50 microinches;
Gold—50 microinches (min.).

We also have requirements that say we have to remain below 220°C temperature. Table 6.2 shows six common solders that are used in many electronic applications.

Table 6.2
Common Solder Characteristics

Composition	Temperature (°C/°F)	Coefficient of Expansion (ppm/°C)	Flux
50 In, 50 Sn	125/257	20	RMA/R
80 In, 15 Pb, 5 Ag	142/290	10	RMA/R
100 In	157/313	29	RMA/R
63 Sn, 37 Pb	183/361	25	RMA/R
60 Sn, 40 Pb	188/370	27	RMA/R
50 In, 50 Pb	209/408	27	RMA/R

As we start our search for the proper solder, we can eliminate three solders immediately because of their tin (Sn) content: numbers one (50In/50Sn), four (63Sn/37Pb), and five (60Sn/40Pb). Our previous discussion explained how tin and gold will form a very brittle and unreliable connection.

Number six (50In/50Pb) would probably be eliminated because its melting temperature (209°C) is approaching the maximum limit imposed by our initial requirements. If you can guarantee that the 220°C figure would not be approached, this solder could be used. Usually, however, you should allow some sort of safety factor when using solders. This solder should be eliminated for this application.

Solder number three (100In) would probably also be eliminated because pure indium solder is very soft and should not be used for attaching substrates to a carrier. There are times when a small chip component (capacitor or resistor) can be attached with 100In, but a substrate has too large an area for pure indium solder to be used.

The one remaining solder, number two (80In/15Pb/5Ag), fulfills all of the necessary requirements to solder over a gold metallized substrate:

- It contains no tin (Sn).
- It has a small content of silver (Ag) to aid in a good gold solder connection.
- It melts at a temperature substantially below the maximum 220°C temperature (142°C). This will not dissolve the gold like higher-temperature solder will.

(Note that this solder also has a coefficient of expansion of from one-half to one-third that of the other solders. This may be a benefit or a liability, depending on the application. Be sure to check this out before using any solder for any application that has to operate over a large temperature range.)

This is one example of solder used to solder to gold. There are some conflicting views on exactly how to accomplish this task. Most papers published on this topic favor indium solder and deal with three concepts concerning soldering to gold: scavenging, wetting, and aging. These terms are very important when choosing a solder for gold applications.

There are also some papers that say that tin-lead can be used to solder to gold if the gold content of the solder connection is kept very low. Most testing shows that this is a critical balance that few people can achieve. For this reason, a good rule to follow is to use solder with *no tin content* when soldering to gold.

Let us backtrack a bit here to look at the three concepts listed earlier: scavenging, wetting, and aging. The first two have been mentioned previously but will be covered in more detail here to show their relation to gold soldering.

Scavenging is the dissolving of metallization by the liquid solder that makes the soldered pieces useless. Tin-lead solders cause significant scavenging on gold. Numerous tests have been run that compare the gold-lead-tin interface and the gold-lead-indium interface. An overwhelming number of these tests show that there is considerable scavenging with the gold-lead-tin combination and little or no scavenging with the gold-lead-indium combination: an excellent reason to stay away from tin-lead solder on gold.

Wetting is the characteristic of liquid solder in contact with a metal part that causes flow and spreading of the liquid until an equilibrium point is reached. Basically, it is the ability of a solder to flow properly to provide a good connection over the entire contact area. This is important in the soldering of gold because a solder that wets very poorly will take more time (and heat) to work properly. This will tend to dissolve the gold, and the final connection will not be the gold but some other base metal. A good wetting solder is therefore the desired type to use.

The aging process is the change in a solder connection that occurs because of a thermodynamic interaction between the connection and the environment to which it is exposed. This interaction causes metallic compounds to grow at the connection. Such intermetallic compounds as $AuIn_2$ and Au_9In_4 form and can impair the mechanical bond of the solder connection. For this reason, indium-lead solders should not be used when:

- Solder is less than 15 mm thick.
- Gold film is thicker than 10 mm.

In this case, the 80In/15Pb/5Ag solder that we previously chose for our example should be used. The 5% Ag (silver) will cut down the development of these intermetallic compounds.

All soldering is not done with a hot iron. Substrates and components are often attached by means of a *reflow* process. In this process, preforms of solder are placed between the device to be soldered (substrate or component) and the ground plane or substrate, and the entire assembly is heated. The solder then flows throughout the open space between the units. This process can be controlled very closely and is a good method for covering larger areas with solder.

Another method that can be used is *vapor-phase soldering.* This is also a reflow process in that it melts a solder that has been previously applied. In this process, the pieces to be soldered are "immersed" in an atmosphere of saturated vapor from a boiling liquid. The vapor from the boiling fluid completely envelops the pieces to be soldered and begins to condense, giving up its latent heat of vaporization. This heat rapidly and uniformly raises the temperature of the pieces to be soldered to the boiling point of the liquid and thus melts the solder. Figure 6.2 shows a basic vapor-phase system. This type of system is good for some application where a bit of soldering is needed.

At this point, you should realize that both tin-lead and indium solders have very definite and acceptable places in microwaves for a variety of solder applications. The trick is to find the right solder for your application, which will provide a solid mechanical and electrical bond that will be reliable for many years of operation.

6.3 Epoxy

When solder is used, you are effecting a metallurgical type of bond; that is, two metals are actually being joined together. When epoxy is used, you are developing an adhesive type of bond. This is the all-purpose glue for electronics and microwaves.

Epoxies are probably the most difficult types of substances for uninformed individuals to understand. Generally, when something is heated to a temperature of 100°C or higher, it becomes a thin liquid and then hardens as it cools. Contrary to this concept, the epoxies actually harden (or *cure*) at these elevated temperatures and remain hard when the circuit is cooled. Many a program manager or components person has argued that a component or substrate must float when the epoxy is cured and cause inconsistent

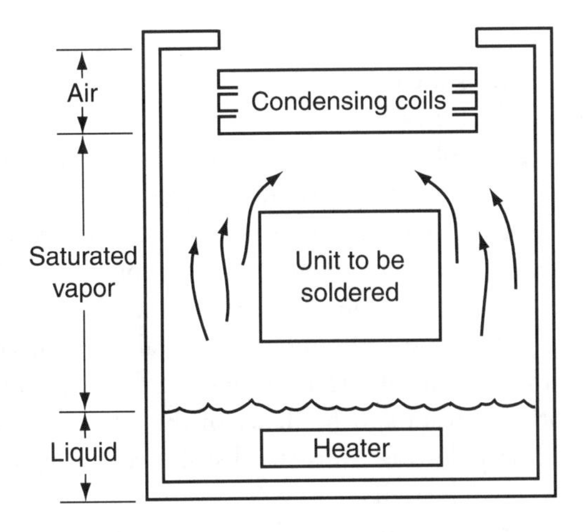

(a) Basic system

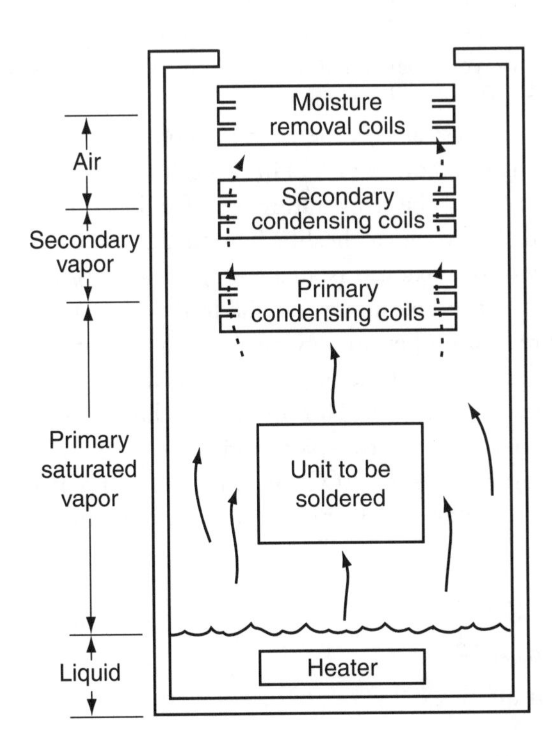

(b) System with secondary vapor blanket

Figure 6.2 Vapor-phase soldering.

connections only to find out that the connection has actually gotten better as the epoxy cured. This is the first concept that must be grasped to fully understand epoxies and their uses.

The first and most obvious question to be asked is: What is epoxy? Epoxy is defined as a thermosetting material used for adhesive purposes. This definition is adjusted to fit the applications used in microwaves because epoxies are also used for potting. We are, however, only interested in them as adhesives in this text. Epoxy comes in two types: one-component (part) and two-component. The one-component type is already mixed, while the two-part requires the mixing of the resin and hardener to form the final epoxy mixture. Both the resin and hardener of the two-part epoxy contain either silver or gold particles that must be distributed through the paste to ensure good electrical conductivity.

To aid in understanding this seemingly unorthodox type of material, we will present frequently used terminology for epoxy and define these terms in understandable language. We will then show typical epoxies that are used in microwaves to help you make the proper decision as to which one may be best for your particular application. Terms to be covered are *cure time, shelf life, pot life,* and *trixotropic.*

Cure time, very basically, is the time the epoxy takes to form a good electrical and mechanical connection. This time is determined by the composition of the epoxy and the temperature used for the curing process. Curing something usually involves a “healing” process, and you could relate this same type of process to epoxies. It is, in a sense, a healing process that brings the epoxy from a silver or gold paste material to the hard and solid electrical connection required for proper microwave circuit operations. You can think of it as being similar to concrete setting up from its flowing form into a solid form. Curing times can run from 10 minutes to 2 hours, depending on the makeup of the epoxy (one- or two-part) and the curing temperature used. (These can range from a low of 80°C for some two-component epoxies up to 260°C for some single-component epoxies.) Typical cure times for two-part epoxies are:

- 1 hour at 150°C;
- 45 seconds at 178°C;
- 3 hours at 80°C;

and for one-part epoxies:

- 1 hour at 180°C;
- 1½ hours at 165°C.

Be sure to check the cure time of the epoxy that you intend to use. It is important that the cure temperature does not exceed the maximum temperature of the components you are going to epoxy (resistors, capacitors, transistors, etc.). You also should check cure temperatures if you are doing two separate operations; that is, epoxying a substrate to a base plate and then epoxying components to a substrate, or vice versa. The first epoxy should have a higher curing temperature than the second one so that the initial connection will not be disturbed when the second curing process is executed.

The second term to be defined is shelf life. This is a length of time between shipment of the epoxy by the manufacturer and a date where you will no longer obtain the proper bonding and conductivity of the epoxy. Typical shelf life for one-part epoxy is about six months when refrigerated and from one to two years for two-part epoxy. One difference between one-component and two-component epoxy is that the shelf life applies at room temperature (25°C) rather than requiring refrigeration to maintain its life. One important point to keep in mind is that the closer the epoxy comes to this shelf-life date, the more questionable its operation will be. Always try to use "fresh" epoxy if at all possible.

Pot life is very similar to shelf life, except that the clock begins when you remove the epoxy from the original container as opposed to when the epoxy is shipped. Pot life is the length of time that the epoxy will be "good" after it is taken from its original shipping container.

Once again, there is a drastic difference between one- and two-component epoxies with respect to pot life. This time, however, the difference is reversed from that of shelf life. Pot life for one-component epoxy ranges from one week to three months, and for most epoxies it ranges from one to two weeks. The pot life for two-component epoxies, however, ranges from two to four *days.* If you are going to use two-component epoxy, you should mix only enough for the job at hand so that it will not be wasted if the pot life is exceeded. Even if you are using one-component epoxy, a good idea is to remove only what you think you will need for the job. This will keep your assembly area from being cluttered with small batches of epoxy that probably should have been thrown out days earlier.

The final term to be defined is trixotropic. This term often appears on epoxy data sheets to describe its consistency. It is a term not usually found in any dictionary. The easiest way to think of the term "trixotropic" is to consider that it is a fluid that thins the epoxy paste so that it can be easily applied. Many thick-film handbooks speak of screening resistive paste, a conductive paste on a substrate, and say that a moderate degree of trixotropy is used to provide good definition of the screened area. They generally continue with a statement that highly trixotropic pastes tend to leave an imprint of the screen mesh; that is, they have thinned the paste too much so that it has more of a watery texture than a smooth, even paste. You can think of the term trixotropic as referring to a process for thinning epoxy so that it is easy to handle.

With the basic definitions used in epoxies presented, we now investigate specific epoxies. As stated previously, conductive epoxy can be either a one-component type or may consist of two parts. The one-component type is premixed and simply needs to be applied to the surfaces to be joined. The two-component type must be mixed in a prescribed manner and volume in order to result in the proper adhesive and conductive properties.

You may question the need for two types of epoxy. The easiest would seem to be the one-component type, where no additional mixing is necessary. In many cases, this is true. There are, however, some disadvantages to the one-component epoxy that are eliminated by the two-component type. To distinguish between epoxies, we will list advantages and disadvantages of each.

One-Component:

- Short shelf life (six months);
- Needs refrigeration;
- Long pot life (two weeks to three months);
- Needs no parts mixing;
- Shorter curing times;
- Higher curing temperatures.

Two-Component:

- Long shelf life (one to two years);
- Needs no refrigeration;
- Short pot life (two to four days);
- Parts must be mixed (resin and hardener);
- Longer curing time;
- Lower curing temperature.

Just as everything else we have covered in this text thus far has trade-offs, so also are there trade-offs to be made when choosing an epoxy. You should evaluate the pros and cons of each epoxy and choose the proper one for your application.

Table 6.3
Common Epoxies

Epoxy Type	Number of Components	Pot Life	Curing (Minutes)	Shelf Life
H20E	2	4 days	90 @ 80°C	1 year
H20E-PFC	2	3 days	90 @ 80°C	1 year
H20E-175	2	4 days	60 @ 150°C	1 year
H35-175MP	1	N/A	90 @ 165°C	¼ year
H37MP	1	N/A	60 @ 150°C	½ year
58C	1	N/A	50 @ 177°C	¼ year

To understand what epoxies are commercially available, we will present one- and two-component epoxies from a variety of manufacturers and list terms that a data sheet would show. Table 6.3 shows epoxies from four vendors—Epoxy Technology, Ablestik Laboratories, Emerson and Cuming, and Amicon (Polymer Product Division)—and lists a wide variety of epoxies with various values of pot life, shelf life, and curing times. One term that is not in this table is *volume resistivity,* which is a measure of the conductance of the epoxy. This value ranges from 0.0001 to 0.0003 Ω-cm for epoxy technology H20E and H20S; 0.0001 to 0.0005 Ω-cm for H40; 0.0005 to 0.0009 Ω-cm for H81; 0.0001 Ω-cm for Ablestik 58-1; 0.00004 Ω-cm for 36-2; less than 0.002 Ω-cm for Emerson and Cuming 58C; less than 0.0005 Ω-cm for Amicon CT 4042-5; and 0.001 Ω-cm for C-860-XCC. Once again, note the wide range.

A final term that should be covered in case the application is for space equipment is *outgassing.* This phenomenon is created by a vacuum environment and causes a material to release vapors or gas when subjected to a vacuum condition that hampers circuit operations. The outgassing should be kept to a minimum in a good epoxy.

These are the terms used to describe conductive epoxy adhesives for microwave applications. As previously mentioned, you should evaluate each material as a separate entity, weigh the pros and cons, and make a decision as to which is best for your application.

With the terminology presented and defined, the next question may be: Where do you use epoxy? There are two application areas for epoxy in microwaves: (1) component attachment and (2) substrate attachment. Both of these areas will be covered in detail later in this chapter. For now, we can say that with the proper conditions set up and precautions observed, you may use epoxy very effectively for attaching microwave components (capacitors, resistors, etc.) to substrates and substrates to carrier plates or cases.

Now that we have become familiar with the terminology and basic applications, our next step is to learn how to use the epoxy. The first item to consider when planning to use epoxy is the *preapplication cleaning.* As in soldering, cleaning is important when making a microwave connection with epoxy. Of prime importance is the substance that you use to do this cleaning. Absolutely *do not use alcohol* or alcohol products. A freon bath from about 23°C to 36°C for less than 5 minutes will do a good cleaning job, as will a brief immersion in trichlorethylene followed by a freon bath.

Chemicals such as toluene or xylene should *not* be used. Toluene has a high degree of toxicity, and xylene has cleaning properties that can be likened to cleaning with kerosene.

Other chemicals or solvents can be used prior to epoxy application. If you have a chemical you would like to use but do not know if it is suitable for use, call and discuss it with the epoxy manufacturer. The vendor usually will discuss methods with you and make recommendations that will reduce, or eliminate, your problems and give you a well-behaved microwave system as an end product.

The next area of importance when working with epoxy is the *mixing.* (This passage does not apply if you use one-component epoxies; they are already mixed for you.) When using two-component epoxies, there is a definite way to combine the individual components (resin with silver or gold powder and hardener with silver or gold powder). Instructions for these epoxies say that they are mixed 1:1 by volume or weight; that is, equal quantities of resin and hardener. There is also an instruction below the mixing ratio on most data sheets that says: "Mix the contents of part A (resin) and part B (hardener) thoroughly before mixing the two together."

This simple one-line statement is vital to achieving a reliable and lasting connection with two-component epoxy. If the two individual components (resin and hardener) are not *thoroughly* mixed *before* the two are mixed together, there will be substantial quantities of silver or gold powder that are not mixed throughout the paste. This causes areas of high resistance in some areas (where there is a low concentration of silver or gold powder) and areas of low resistance (high concentration of powder) in others. What we seek, ideally, is a material that has a high degree of consistency throughout the whole area. This can be accomplished only if each individual part is mixed thoroughly prior to mixing the two components together. Whenever you have to use two-component epoxy or have someone else apply it for you, be sure that it is mixed properly.

With the epoxy mixed, the next step is to apply it to the surfaces that are to be joined. Epoxy has been applied in a variety of ways over the years, with everything from screens to needles to toothpicks to the ultimate automatic machine. The screen and the needle are the most widely used types of applicator. A 200-mesh screen is usually used with approximately a 0.002″ layer of epoxy deposital. This would be the method used when you are epoxying down a substrate to a carrier plate or case. It is much more feasible when a large area is being epoxied.

A fine needle should be used to attach components with epoxy. This will dispense a precise dot of epoxy in the area of connection; this dot should usually not exceed 0.005″ in diameter. The component is then placed on top of the epoxy dot and pressed into place. Care should be taken not to press too hard, as the epoxy can flow out from under the leads and possibly short a component to itself or to ground. A gentle pressing is sufficient.

To this point, we have spoken of joining substrates to carrier plates, substrates to base plates, and components to substrates without much consideration as to what material is being joined together or what effect a certain material may have on the epoxy being used. As in the solders, there are some metals that are just not compatible; for example, silver on aluminum, tin, or lead surfaces. You should *not* use silver epoxy on any of these surfaces (that is, on bare aluminum or a tin- or lead-plated surface); it results in no real bond between the surfaces and a brittle connection. This, in turn, results in a high-resistance connection that eventually cracks and separates (especially under environmental conditions). Gold epoxy also should not be used on bare aluminum for the same reasons. You should make a habit of using only screws to attach any substrate to bare aluminum.

Some people try to use an iridite (alodine) process that puts a coating over the bare aluminum. This process should not be used where high-reliability applications are involved or where a ground plane requiring high conductivity (which is always the case in microwaves) is needed. Whenever aluminum is involved, have it gold- or copper-plated to avoid immediate and long-term problems. This may sound expensive, but sometimes you can plate the base plate only and apply an iridite to the side walls of the case and still have a circuit that works very well. Other times, however, a completely plated case is a necessity. This decision will depend, once again, on your application.

To summarize our discussions on epoxies, we will list the qualities that make an epoxy suitable and acceptable for microwave applications:

- Low-volume resistivity (0.0001 to 0.0003 Ω-cm);
- Lap shear strength between 1,000 and 2,000 psi;

- Flowing properties for screening ease;
- Pot-life stability (uniformity in the batch);
- Good shelf life (at least six months);
- Minimum outgassing;
- Minimum bleeding (or spreading) of the resin;
- Curing temperature compatible with the substrate or component being used.

With these requirements met, you will have a very acceptable adhesive for microwave applications.

One statement appears on all epoxy data sheets that has not yet been mentioned: a *caution* statement that reads something like this: "This product may cause skin irritation to sensitive personnel. If contact with skin occurs, wash the affected area immediately with soap and water." Pay close attention to any cautions printed on the data sheets, and observe them to the letter. Careful attention in the beginning will pay off later when accidents or painful experiences are avoided.

6.4 Bonds

This chapter first discussed soldering and classified it as a metallurgical bonding process, and next covered epoxies and called this adhesive bonding. We will now go back to a metallurgical type of bonding and discuss thermocompression, ultrasonic, and thermosonic bonding techniques. Each of these methods is used for attaching microwave components to substrates. These are specialized types of bonds because they are not used to attach large areas such as substrates to carriers, as are soldering and epoxy methods; rather, these are wire-bonding methods. They are designed to attach one area to another (usually from a chip to a connecting pad or a circuit). The bond is formed by placing a wire (or strap in some cases) on the appropriate metal pad and having a bonding tool with heat and pressure placed on this wire. The heat and pressure actually deform the wire (or strap) material and force it to be bonded to the metal pad. This bond comes about by the interaction of atomic forces between the wire and metal. The actual metal bonding process consists of three variables:

- *Pressure* or *force*—to promote a plastic flow and close contact between metals;

- *Elevated temperature* (or *heat*)—to promote contaminant dispersal while lowering flow stress and improving the diffusion process;
- *Time*—to promote solid-state diffusion in the actual bond zone.

This process should not be confused with a commonly known term, *eutectic bonding.* A eutectic bond also involves heat and pressure, just as described above, but this type of bond requires a third metal film to be placed between the wire and metal pad. This produces a diffusion process when heat and pressure are applied. The bonds we are describing above, and throughout this section, are brought about by interatomic forces.

In the wire bonding process, the wire itself must be ductile (be able to be drawn thin without breaking) and must be capable of deforming to comply with the bonding tools used. The metals that are most widely used in this process are gold (Au) and aluminum (Al), with aluminum being preferred in many applications because of its more rigid structure once the bond wires have been put in place. Diameters of 0.7 mil (0.0007″) to 1 or 2 mils (0.001 to 0.002″) are common wires used for bonding.

The metal to which the wires are bonded must have a composition that does not have a high degree of oxidation. Materials such as copper (Cu) or palladium (Pd) are not generally used because they are very susceptible to oxidation. Materials such as platinum (Pt), gold (Au), or silver (Ag) are used extensively, however, because they have very little tendency to oxidize. You can see how an oxidation on the surface of the metal could obstruct the bonding process because it would be like an impurity film on top of the surface that would be between the two metals that were attempting to join. This would be much like not cleaning a circuit board prior to soldering or applying epoxy. The bond may take place initially, but its integrity and reliability would be questionable. We will now cover the three types of bonding processes listed previously: (1) thermocompression, (2) ultrasonic, and (3) thermosonic.

6.4.1 Thermocompression Bonding

Thermocompression bonding is almost self-explanatory. The two parts of the name say just what is involved in the bonding process: *thermo* means a temperature, and *compression* relates to pressure. Thus, thermocompression is accomplished by combining heat and pressure to form the interatomic structure needed for wire bonding. The heat is applied either to the substrate, the chip being bonded, or both, while pressure is applied to the bond area. In this way, both the temperature and pressure can be very closely controlled and applied to the appropriate area.

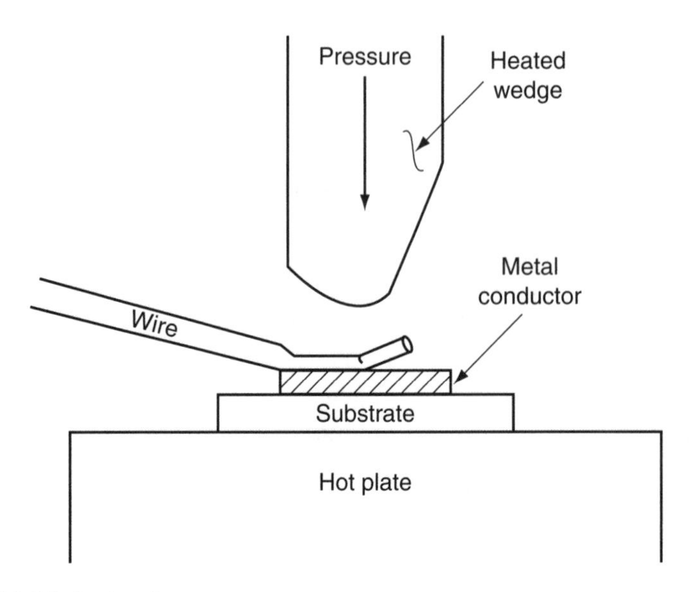

Figure 6.3 Wedge bonding.

There are three basic types of thermocompression bonds: the *wedge bond*, the *ball bond*, and the *stitch bond*. Each of these has specific areas where it should be used. The type you choose, once again, depends on the particular application, substrate temperature, or pressure limitations, and the metals involved in the bonding area.

Wedge bonding is shown in Figure 6.3. You will note that both the wedge and the substrate that is being bonded to are heated. This process, combined with the pressure applied by the wedge tool, provides an excellent connection. The wedge tool usually is made of a very fine sapphire or silicon carbide.

Bonding occurs when the tool and substrate are at the proper temperature and the wire is in place. The wedge is then brought down for bonding, and the tool deforms the wire to form a highly reliable bond. After a specified time, the tool is lifted, and the bond is complete. This type of bonding is rather like pressing a thin candle on a sheet of paper with a soldering iron: the combination of heat, pressure, and time have caused the wax to adhere to the paper. Similarly, the combination of heat, pressure, and time (all closely controlled) have bonded the wire to the metal conductor. Wedge bonding is used many times when very fine wire (less than 0.7 mil) is to be bonded.

The values used for pressure, temperature, and time will vary from device to device, and the manual for the particular bonder being used should be consulted. As a representative example, the following values are used to wedge bond to a transistor chip:

Temperature—	300°C ± 10° for silicon bipolar transistors, 260°C ± 10° for GaAs field-effect transistors (FETs);
Pressure—	30 grams (40 grams, max.);
Time—	2 to 3 seconds.

These values are examples. Many times trial and error methods are needed to find the ideal bond for your application. This ideal bond usually is when the "footprint" produced by the bonding tool (the depression in the wire) is two to three times longer than the diameter of the wire. This is shown in Figure 6.4. This configuration will result in a good, solid bond on which you can rely to hold for many years of operation.

Ball bonding, also termed *nail head* bonding, uses thermocompression techniques like the wedge bond but does so through a capillary tube. This is a tube with a very small opening. This type of bonding, however, usually is not used for wires as small as 0.7 mils because these fine wires do not flow smoothly through the capillary tubes. Generally, wire of 1 mil or larger is used for ball bonding.

Figure 6.5 shows ball bonding. The bond is formed by first feeding a gold wire through the capillary and forming a ball on the end of the wire with a hydrogen flame. We specified gold wire because this is the only wire that will form the required ball. Aluminum wire cannot be used for ball

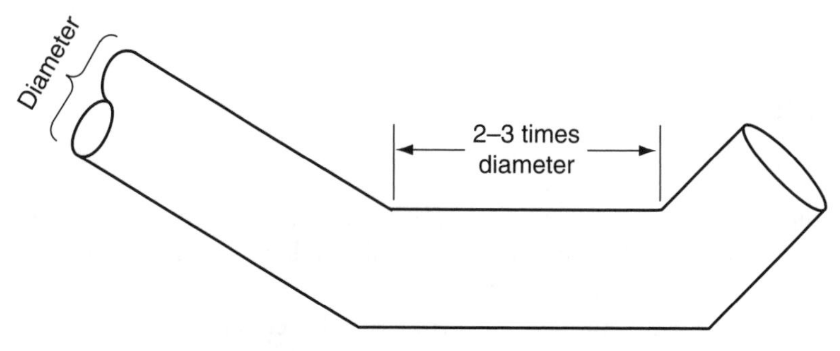

Figure 6.4 Ideal wedge bonding.

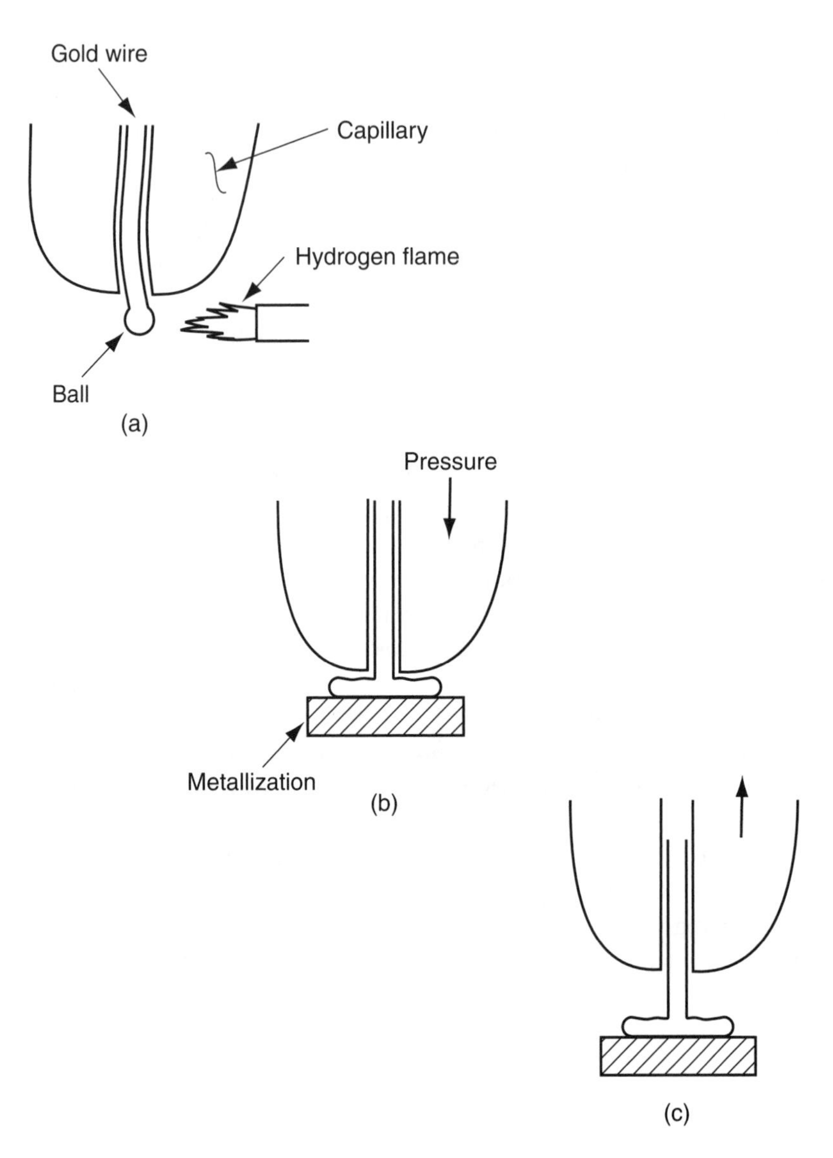

Figure 6.5 Ball bonding.

bonding as it was for wedge bonding because aluminum will not form a ball when cut by the hydrogen flame. (The ball produced by the hydrogen cutting flame is shown in Figure 6.5(a).)

After the ball is formed, the capillary is lowered until contact is made with the metallization. It is then lowered further, where a predetermined force deforms the ball and causes the bond to be formed (Figure 6.5(b)). In this condition, the substrate or the capillary is heated. Usually, the heated-probe-tip (capillary) type of bonder is preferred because the substrate does not need to be heated and there is no danger of destroying or damaging a component to be bonded. With this type of bonder, the heat is generated by sending a pulse of current through the tip of the tool while the bond is being made. This means that the device is heated only in the immediate area of the bond being made. After a specified time, the capillary is raised from the substrate. This leaves the "nail head" bond to which we previously referred (Figure 6.5(c)). The hydrogen flame once again heats the wire, cuts it, and forms a new ball for the next bond.

The third type of bond is the stitch bond, shown in Figure 6.6. The stitch bond is a combination of the wedge and ball bonds. It is like the wedge bond because the wire is depressed to make contact with the surface to be bonded, and it is like the ball bond because a capillary is used. The stitch bond is begun by feeding the bond wire (either gold or aluminum) through the capillary tube and bending it at a 90° angle. This bend, and the absence of a ball, allows the use of either gold or aluminum wires, as previously mentioned. The initial bend is shown in Figure 6.6(a).

The next step in a stitch bond is to lower the heated capillary and apply pressure to the bond wire to deform it and form the desired bond. (The substrate may also be heated, but you should be careful that the temperature used does not degrade the parameters of the circuit.) The deformation of the bond wire is shown in Figure 6.6(b).

In Figure 6.6(c), the bond has been made, and the capillary is pulled back to begin making a second bond. In Figure 6.6(d), the bond wire is cut by metal cutters rather than the hydrogen flame bonding. This, once again, allows you to use either gold or aluminum wire for your bonds.

One very important point must be brought up at this juncture: the bonding tool must be moved *in the direction of the stitch* when going from the first bond (Figure 6.6(b)) to the second (Figure 6.6(c)). If it is not moved in this direction, a stress point is created directly at the bond that more than likely will cause a bond failure. It is vital, therefore, to position the substrate so that the first and second bonds are directly in line. This can be considered to be a disadvantage of the stitch bond in that it must be so critically positioned.

These are the bonding methods that are available when using thermocompression bonding. In review, the following requirements should be met to ensure a good and reliable bond:

- Choose the appropriate wire material (gold or aluminum).
- Be sure that sufficient deformation of the wire occurs when making the bond. The bond length should be from two to three times the diameter of the wire.
- Choose, and optimize, the time, pressure, and temperature to make the best bond for your application.
- Be sure that the bonding surface is clean.

If these procedures are followed, you should have no trouble with your bonds. You should, however, be aware of some of the causes of poor

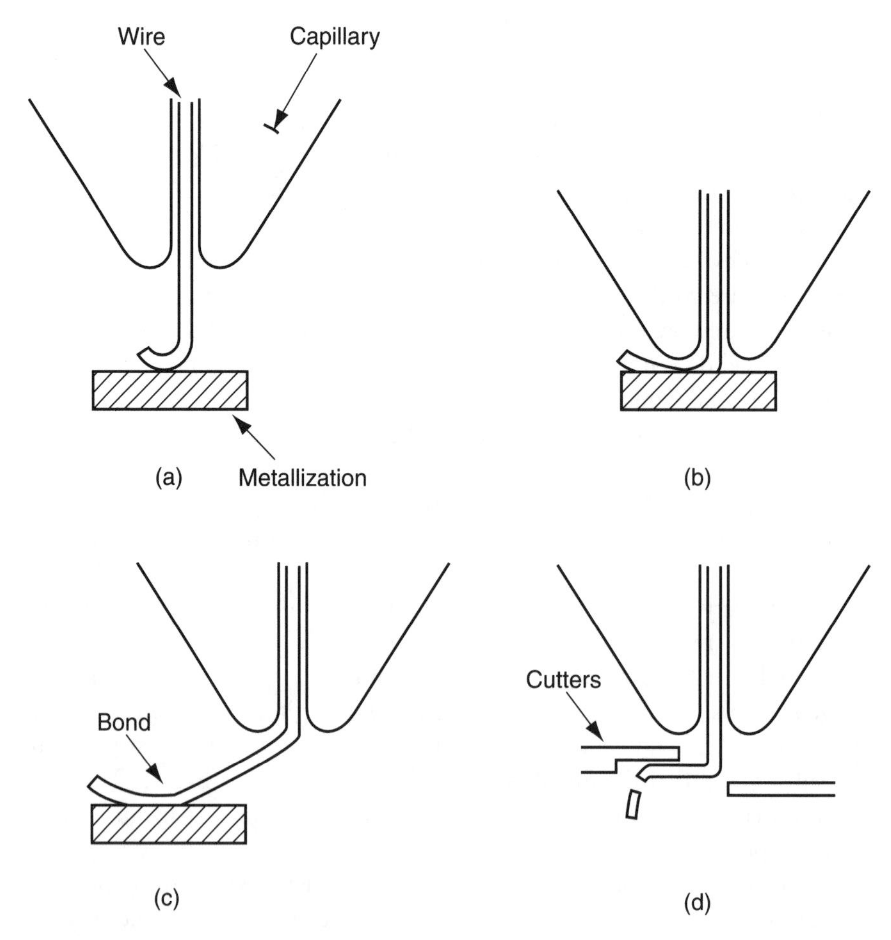

Figure 6.6 Stitch bonding.

thermocompression bonds. The three primary causes of poor bonds and a brief explanation of each are listed here:

- *Too much or too little heat or pressure*—These conditions can be detected by examination under a microscope. They will show up as a bond wire that has either very little deformation or excessive deformation. Remember the number of two to three times the wire diameter quoted previously for an ideal bond.
- *Impure wire*—This will show up as a bond wire that will not deform properly. Even when increased pressure and heat are applied, it will not deform properly. This is because even the smallest amount of impurity within the wire will cause it to harden and will not allow it to deform to produce the proper bond.
- *Dirty or glassy metallization*—This will cause a properly deformed bond wire either not to bond at all or to form a very weak bond. All oil, dirt, and oxides must be removed from the metallization before you attempt any bonding process.

One additional problem that can arise when bonding aluminum to gold is what is called the *purple plaque.* This is a brittle alloy that forms when an aluminum-gold (AlAu) bond is exposed to extended heat. This is why both the temperature and duration are so important when bonding aluminum and gold. This problem causes frequent bond failure because the gold is absorbed into the newly formed alloy, and adhesion between the wire and the metallization is lost. Take great care when choosing a bonding temperature *and* duration.

6.4.2 Ultrasonic Bonding

Ultrasonic bonding is a *cold* bonding process; that is, there is no heat associated with the joining of two metals other than that set up by a scrubbing action of the bonder. This type of bonding has many advantages where delicate circuitry should not be exposed to elevated temperatures that could be associated with other bonding techniques.

Figure 6.7 shows an ultrasonic bonding setup. You can see that it looks much like all of the other bonding methods that use a capillary tube but has two exceptions. There is an ultrasonic head attached to the capillary tube, and there is absolutely no reference to temperature on either the substrate or the capillary tube. This is because the process uses a scrubbing

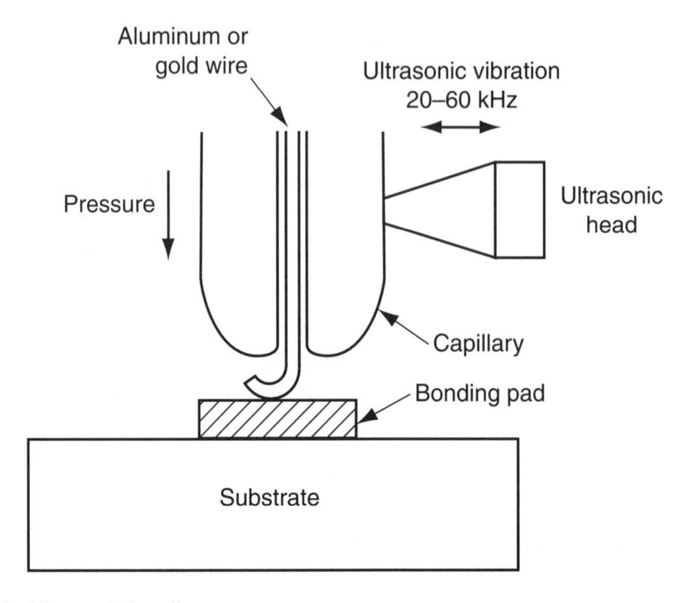

Figure 6.7 Ultrasonic bonding.

action and pressure to join the two metals, and no heat is needed to deform the wire to form the bond.

Ultrasonic bonders can use both aluminum (Al) and gold (Au) wires as well as silver (Ag) and copper (Cu). Both ball and stitch bonds can be made with the same restriction placed on ball bonding as in thermocompression bonding: aluminum wire cannot be used because of its inability to form the required ball when cut by the hydrogen flame.

The advantages of ultrasonic bonding are many and make this process one that should be considered for production application:

- Requires no external heating;
- Has rapid bond rates;
- Is low cost;
- Allows easy replacement of chips without damage.

As with everything, the ultrasonic bonding process also has disadvantages:

- Settings on the bonds are very critical.
- The ultrasonic oscillator will drift if not left on to stabilize.

- Rough and uneven bonds can occur.
- Alignment is more difficult with this process.

If the disadvantages listed can be overcome or minimized, the ultrasonic bonding process is one that can find many applications. One further point should be mentioned before leaving the discussion on ultrasonic bonding: temperature. We have said that no external heat is required for the process, and this is true. We also said that the only heat present was from the scrubbing action of the capillary at an ultrasonic rate. You should understand that this scrubbing action can cause a temperature equal to 30–50% of the melting point of the metals being joined. Do not reach for the material following an ultrasonic bonding process, or you may be looking for a way to cool your hand rather quickly. This is an important point to bring up because, even without an application of external heat to the material, there is a rise in temperature that can be considerable.

6.4.3 Thermosonic Bonding

Thermosonic bonding takes the best of the worlds of thermocompression and ultrasonic bonding to produce a process that is finding wide application in microwaves today. This type of bonding is also termed *hot work ultrasonic.*

As mentioned, thermosonic bonding is a combination of thermocompression and ultrasonic bonding and as such results in a process that requires lower bonding temperatures and lower bonding forces. You will recall from the section on thermocompression bonding that typical values for bonding to a transistor chip were a temperature of 300°C for a bipolar device (260°C for GaAs FET); pressure of 30 grams (40 grams max.); and duration of 2 to 3 seconds. With thermosonic bonding, these figures can fall into the range of 150°C temperature; a pressure of approximately 20 grams; and still have a duration in the neighborhood of 2 seconds. We can have reduced temperatures and pressures because the thermocompression mode of the bonding process can operate at a lower temperature (150°C) with the ultrasonic scrubbing energy making up the difference for temperature and bonding pressure; that is, the bonding area is heated just high enough with the capillary to begin a bonding process. Then, the ultrasonic movement of the capillary completes the process, resulting in a lower temperature and pressure.

Bonds produced with thermosonic bonding are the ball bond and the stitch bond. (Remember, once again, that aluminum wire cannot be used for ball bonding in the thermosonic process.) These bonds are fabricated exactly as described previously and will exhibit the same properties as discussed.

6.5 Component and Substrate Attachment

With the various bonding methods now presented, we can see how these methods apply to actual microwave applications. The material presented here will be representative examples and recommendations for attaching active and passive components and substrates. For specific procedures, you should consult the component or substrate manufacturer for directions for your particular application.

There are basically two methods of attaching a substrate to a carrier plate or case: solder and epoxy. Solder is usually applied by a reflow process so that a uniform coating can be achieved. Epoxy should be applied by a screening process so that this same uniformity can be achieved. This uniformity is of prime importance in the attachment of substrates because the bond not only is holding the substrate in place but is providing a *ground plane* for the circuit. Be sure that a uniform flow of solder or thickness of epoxy is applied to attach a substrate. These techniques are valid for PTFE/glass, high-dielectric PTFE laminates, and alumina substrates. The point to remember is one that was brought up before: do not attach any substrate (solder or epoxy) to an unfinished aluminum plate or chassis. Contact will be inferior and the ground plane will not be continuous with this arrangement. Also, be aware of any intermetallic conflicts such as the gold-tin problems listed in previous sections. These can cause brittle and unreliable connections that may appear adequate when first made but will deteriorate rapidly with time and environmental conditions. Be sure to consider all possibilities when deciding which material to use to attach your substrates.

Passive components, such as chip capacitors and resistors, can be attached to a substrate in a variety of ways. The method used, of course, depends on the terminations (connection points) on each individual component. Chip resistors, for example, are available with platinum-gold, silver over nickel, or solder terminations (the solder terminations are optional and must be specified separately). Both of the standard terminations for chip resistors can be attached with the same type of solder or epoxy:

- *Solder*—Reflow process using 60/40 or 63/37 tin-lead solder. If an iron is used, you should not exceed 35 watts with a small chisel point.
- *Epoxy*—Silver-filled conductive epoxy is preferred with a curing temperature *below* 150°C.

Chip capacitors are somewhat more involved because they are available with a variety of terminations. Typical combinations are:

- Platinum gold;
- Gold over chromium;
- Platinum silver;
- Gold over nickel.

Attachment for each of these terminations is as follows:

- *Platinum gold*—Solder with SN60, SN62, or Indalloy 7 (50% indium, 50% lead); eutectic bonding; or any standard gold- or silver-filled epoxy. (*Note:* Check the curing temperature of the epoxy you choose, and be sure it does not exceed that of the capacitor.)
- *Gold over chromium*—Thermocompression, ultrasonic, or thermosonic gold wire or aluminum silicon wire bonding; eutectic bonding; and standard gold- or silver-filled conductive epoxies with the same temperature restrictions for curing as mentioned above. Special solders can also be used that have a low tin content or no tin at all. (Recall the tin-gold reaction mentioned previously.)
- *Platinum silver*—Solders such as SN60, SN62, and Indalloy 7 (50% indium, 50% lead); standard gold- or silver-filled conductive epoxies.
- *Gold over nickel*—Thermocompression, ultrasonic, and thermosonic gold wire or aluminum silicon wire bonding; special solders that have low tin content or no tin; standard gold- or silver-filled epoxies.

A passive device that requires wire bonding for its operation is the Lange coupler. Figure 6.8 shows a drawing of the coupler and where the wire bonds are to be placed. The bonds used in this particular case are wedge bonds.

When considering a method of attachment to a circuit, remember that active devices require more care than passive devices. You need to be concerned not only with intermetallic interactions but also with the temperature used for attachment, since too high a temperature can damage sensitive devices. Such components as diodes and transistors must therefore be watched very carefully.

Table 6.4 shows attachment methods for microwave diodes on a variety of boards. The term *die-down* refers to the putting of the chip into a circuit. *Strip-down* is with the attaching terminals under the diode, and *top*

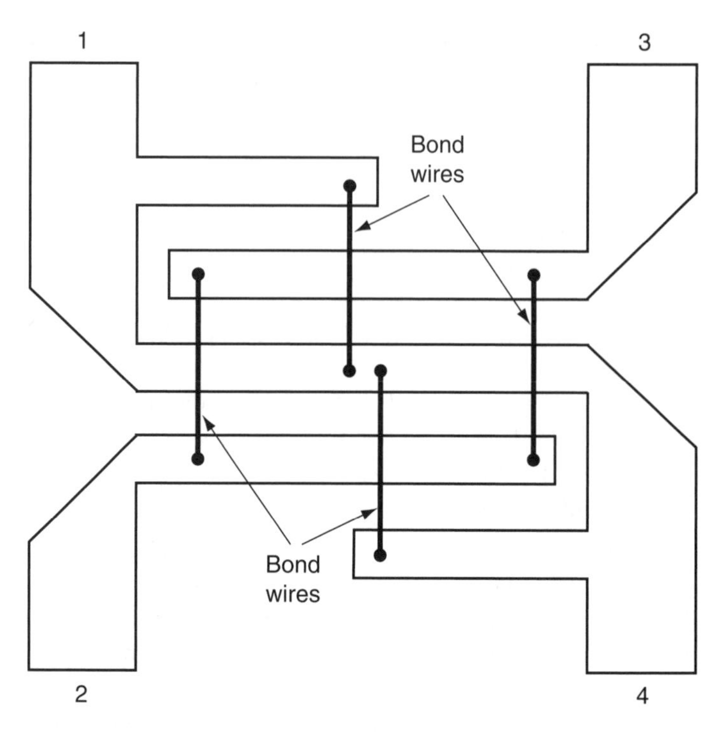

Figure 6.8 Lange coupler.

contact is with the terminals on top of the device necessitating a strap connection.

Keep in mind that the methods shown in Table 6.4 are recommendations and are usually the best methods to use. You should, however, consult the individual data sheets or manufacturer if you have any questions as to which is the best method for your particular diode and application.

Transistor attachment is similar to that of microwave diodes. The major differences in transistor attachment, however, come about from your choice of device: silicon bipolar or GaAs FET. For example, consider the following typical method of die attachment for a silicon bipolar transistor first and a GaAs FET second (both are done in an inert atmosphere).

Silicon Bipolar:

1. Heat to 400°C ± 10° the heater block.
2. Place the circuit that will hold the transistor on the heater block, and allow it time to heat (usually 5 to 15 seconds will do).

Table 6.4
Attachment Methods for Microwave Diodes

Type of Board	Chips With or Without Leads: Usable Die-Down Methods	Chips With or Without Leads: Best Method	Chips With or Without Leads: Best All-Around Method	Beam Leads: Die-Down Method	Beam Leads: Tools Required	Stripline: Strip-Down Method	Stripline: Top Contact
PTFE/glass	1. Soft solder	Hot gas bonder	Soft solder (180° max.)	Not advised to use		Soft solder (180° max.)	Parallel strap welding wire
	2. Conductive epoxy	Epoxy machine					
	3. Eutectic solder (usually)	Hot gas bonder					
Ceramic	1. Eutectic solder	Hot gas bonder	Soft solder or eutectic solder (280° max.)	Thermal compression wedge bond or parallel gap weld	Beam-lead bonding tool	Soft solder	
	2. Soft solder	Hot gas bonder					
	3. Epoxy	Epoxy machine					

3. Place the chip (using tweezers) carefully on the circuit, and scrub back and forth until wetting occurs.
4. Following wetting, do one circular scrub of the chip, and carefully remove the tweezers. A eutectic bond has now been formed.
5. Remove the entire circuit from the heater block and allow it to cool naturally.

GaAs FET:

1. Heat to 300°C ± 10° the heater block.
2. Place the circuit on the block and allow it to heat thoroughly (5 to 15 seconds).
3. Place a gold-tin (AuSn) preform on the circuit in the proper area.
4. Using tweezers, place the chip in the proper orientation on the circuit and scrub it back and forth until wetting occurs.
5. When wetting occurs, make one circular scrub of the chip, and remove the tweezers; this is now a solder bond.
6. Remove the circuit from the heater block, and allow it to air cool.

You now should be able to see the differences between silicon bipolar (a eutectic bond) and GaAs FETs (a solder bond) regarding bonding the transistor chip.

After the chip is in place, the next step is to go from the transistor to the external circuit around it. Wire bonds are used for this purpose—either a gold ball bond or thermocompression (wedge) bond. For higher frequencies, the wedge bond is preferable because of its overall shorter length, which reduces electrical parasitics. The bonds are made as outlined in the previous sections, and specific settings and operations should be obtained from the individual bonder instruction manual. (*Note:* If you are required for one reason or another to use aluminum wire for bonding, be sure to keep close watch on the bonding duration to avoid creation of the purple plaque, to which we referred previously. You will recall that this can take place at an aluminum-gold interface and degrade the bond greatly.)

The use of *surface mount technology* (SMT) has brought about many different problems in the attachment of components to circuit boards. Previously, there was an attachment of a component at the terminations of the resistor or capacitor. This connection generally was at the ends of the components. This is the case for some surface mount components and the same type of attachment schemes should be used.

In other cases, the termination (or connection point) on a surface mount component is underneath the device to be attached. This simply means that the connection must be made in the same way that a resistor, for example, is attached if the termination is on one surface of the device. If the terminations for a resistor are such that the terminations are both underneath the resistor when it is placed on the circuit board, the resistor must be attached by using either a flow solder technique or conducting epoxy. These are the same attachment methods that must be used for a surface mount device that has terminals on its lower side.

So, we can see that the methods of attaching the surface mount components to the circuit are exactly the same as for any component that has been covered in the previous sections. We should also note that bonding techniques, such as those discussed in Section 6.4, can be used as well if there is an area where the bonding wires can be attached to both the circuit and the surface mounted component.

6.6 Chapter Summary

In this chapter, we discussed a variety of bonding methods that can be used in microwave circuits. The very familiar soldering was first presented with the three main areas that are required for a good solder connection: solder, flux, and basis metals.

We then followed with a look at the world of epoxies and how they can be used in microwave circuits. Both one-part and two-part epoxies were presented.

We then finished the chapter with information on thermocompression, ultrasonic, and thermosonic bonding followed by a summary of component and substrate attachments.

It can be clearly seen that a thorough understanding of bonding techniques is required if you are to end up with a microwave circuit that will fulfill the expectations you had when you designed it.

We conclude by cautioning you to treat each attachment in microwaves as an individual and unique case. Analyze what you are trying to bond, take inventory of what materials and methods you have available to you, and you should have a circuit with reliable and dependable connections.

7

Microwave Packaging

7.1 Introduction

The first portion of a microwave circuit that is seen is the package. If you are looking at a group of packages that are pleasantly wrapped and choose the one that looks the best, you are only looking at the surface. When looking at microwave packaging, however, it is a much different story. The package does not need to look good. It needs to be functional and allow the circuit that it contains to work as it was originally designed.

To understand how important the package is in microwave applications, refer to Figure 7.1. The figure shows three methods used for microwave transmission: microstrip, stripline, and suspended substrate. In each case, you see a dielectric (ε), a conductor, and a shaded area that is a ground plane. The last area mentioned (the ground plane) makes the difference in microwave packaging. You may have seen some or all of these configurations and referred to the shaded area as a ground plane (as it is), thinking of it as a piece of copper bonded to a laminate or a block of aluminum used as a mount for your circuit. This block may also have seemed to be a convenient place to mount SMA connectors. This convenient shaded area, however, is much more than a connector or laminate support.

Refer once again to Figure 7.1, and notice differences in how the ground plane is oriented. The microstrip has a ground plane supporting the dielectric with air on top of the circuit; the stripline configuration has ground planes both above and below the circuit; and the suspended substrate

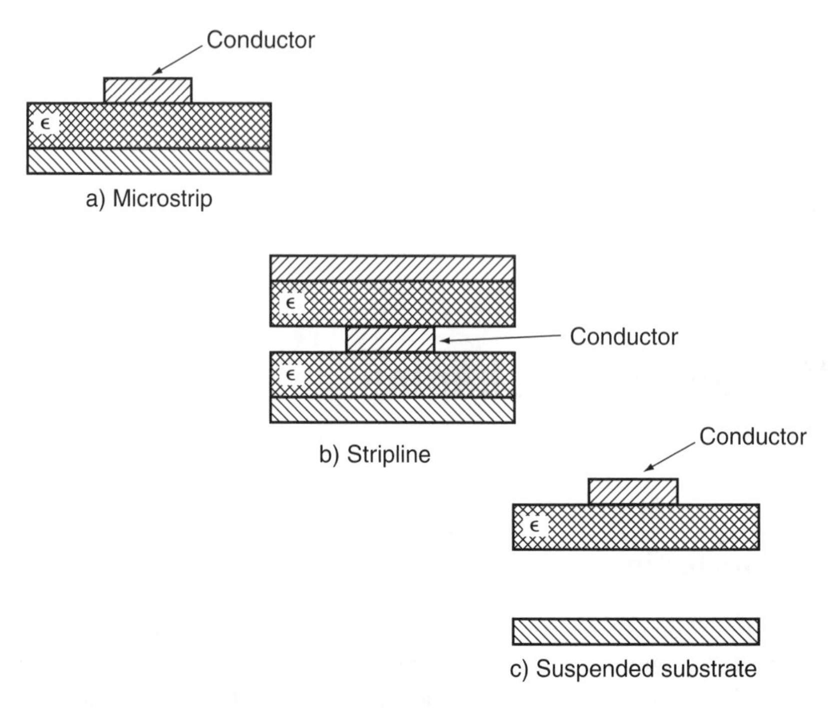

Figure 7.1 Microwave transmission media.

has a ground plane on the bottom of the dielectric, just as in the microstrip case but with an air gap between the dielectric and the ground plane.

The transmission media given in Figure 7.1 are representations of each configuration, showing the orientation of individual parts. Let us put each of them in a package and show what each would look like under final packaged conditions. These are shown in Figure 7.2.

Figure 7.2(a) is packaged microstrip. Notice that the dielectric thickness is designated by b and the dielectric is placed directly on the case floor. The dimension H is the distance from the dielectric and the top of the case. This is a critical dimension and will be covered in detail later in this chapter.

Figure 7.2(b) is packaged stripline. This is the basic stripline package shown in Figure 7.1 with the two ground planes connected together to form a case. The b dimensions are those of the dielectric. You can see how this configuration forms a well-shielded circuit with dielectric and case completely surrounding the circuitry. Various methods of packaging stripline will be covered in this chapter.

Figure 7.2(c) is packaged suspended substrate. This is a method used in millimeter-wave applications. We will not cover this method of transmission medium in any great detail, as it is for higher frequencies and also

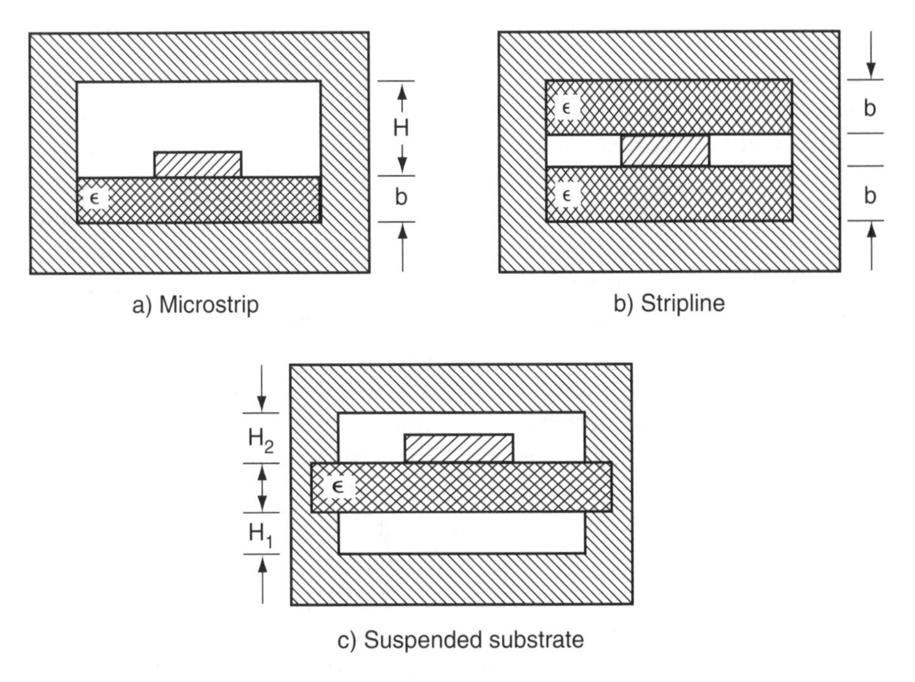

Figure 7.2 Microwave transmission media in cases.

resembles microstrip in many ways. Typical dimensions for suspended substrate are as follows:

Dielectric thickness—0.005″
Dimension H_1—0.010″
Dimension H_2—0.010″
Dielectric width—0.046″

The dimensions are relatively small and indicate higher-frequency operation. Circuits in the 20, 30, and 40 GHz ranges are often fabricated in suspended substrate.

As we have stated previously, the ground plane is an important part of a microwave package. This can be likened to a dc power supply and its ground. If you design an amplifier, for example, to be used in your home stereo system, which will operate on +12 volts, and then put this amplifier in a case, apply +12 volts to it, and forget to put on the ground wire, you will not have the performance that you expect from the amplifier. This analogy is the same for microwave circuits, except that we are not talking about ground wires, but rather ground *planes*. You must understand that when we discuss microwave circuits, we are relating to *waves* propagating down a transmission medium. These waves have more than one component to them in contrast to a dc voltage, which has only one component and thus needs just one ground reference. The electromagnetic wave in microwave signals needs a constant and continuous ground reference in order to have the circuit perform as expected and as designed. This means that the material on which the circuit is etched must always remain in contact with the ground of the case into which it is being placed. This is not always an easy task to perform. There are times when the geometry of the surrounding architecture is such that the ground planes are at two different levels. In this case, you must place a strong emphasis on a strapping arrangement, which will allow the two, or more, levels to appear as a consistent connection for the wave that is being transmitted in the circuit.

Also, because of tolerances and imperfections in board material, there will be times when there is insufficient ground connection to the ground plane by the material with the circuit on it. This *warpage* of the material often causes inadequate ground connection at the connector or launch areas of the cases used for the microwave circuit. This is generally much more of a problem when relatively large circuit boards are used. For small circuit boards, this usually is not a consideration, although the potential problem should be considered for every circuit board.

We mentioned tolerances in our previous discussion. These are the tolerances placed on the circuit material (laminates and substrates) and on the metallic case. Recall that there was a parameter in the alumina substrate data sheet that was called "camber," defined as the deviation from flatness of the material. This parameter is used in the fabrication of cases for microwave circuits. Also, the case itself can be a problem. If the material is basically flat over its entire width and length, but the case has variations in the base plate to which the material is attached, there will also be a grounding problem with the laminate-case combination.

One way to eliminate the problem of inconsistent ground planes is to have the material purchased with a metal plate bonded to it. We have mentioned this material previously and explained its construction and some of its advantages. We will now look at the material from a packaging standpoint. As such, we need to look at the types of material that we use for this backing, why we use it, and how we make ground connections to the circuit when the grounds are not at the edge of the board.

All woven and nonwoven glass substrates can be bonded to a thick metal. This may be a direct fusion bond between the metal and the material or the use of a sheet of adhesive, such as FEP, between the two. Whichever way is used, there is an excellent bond between the material and metal that virtually eliminates any grounding problems.

The most common types of metals for this purpose are aluminum, copper, and brass. The most common aluminum alloy is 6061T6, which is readily available and easily worked. The most common copper used is electrolytic tough pitch copper 110. This is a hard copper that machines well. It has the best thermal properties of the three metals used and can be easily plated if necessary. This copper also is higher in cost than aluminum and has a greater weight. The brass usually used is cartridge brass 250. It has approximately the same properties as copper but is much easier to machine.

We have looked at some of the properties of the metal-backed substrate, but we ask: What are the real advantages of having a material that costs more and seems to be heavier than a standard material? The main advantages of such a system are:

- Dimensional stability and flatness, which keep the material rigid before and after etching. Many soft substrates will "curl" after etching because so much copper has been removed from one side of the board and the stresses on the material are not now the same. This contributes greatly to the maintenance of the flat and continuous ground plane for the circuit. The metal-backed material does not have this problem.
- The metal backing on the material can also serve as a very efficient heat sink for higher-power applications.
- The metal-backed substrate can offer significantly reduced assembly times. The base of the case is already in place, and there does not need to be any thought given to how you will attach the substrate to the case because it is already attached. Only a machining operation is needed.

We have mentioned many times the need to have a good ground plane in microwave circuits. There are, however, times when there must be places within the circuit area where a ground connection can be made. In a conventional circuit board material, a hole or slot is usually cut or milled and a piece of copper is attached to the ground plane. Not readily apparent is how this can be accomplished when the material is attached to a sheet of aluminum, such as the type we have been discussing.

There are basically three methods used to attach an area on a circuit board to the ground plane: *roll pins, press fit pins,* and *plated-through-holes.* The roll pins have been used since the metal-backed material was introduced. These pins are standard split-steel roll pins that are plated for solderability. The pins are squeezed together and released after insertion so that continuous contact is made. One disadvantage is that the pin must extend above the surface of the circuit. The typical method of attachment is soldering by using solder preforms. Conductive epoxy may also be used. Because the pins rely on spring tension for proper connections, solder, flux, and epoxy must be kept from the inside of the roll pin.

The press fit pins are solid and are typically machined from brass that is tin-plated so that it can be soldered. The pins are designed to be inserted with an interface fit into the holes. This type of pin can be inserted flush with the surface of the circuit, which allows the use of surface mount technology. With SMT, solder paste can be screened over the surface of the pin at the same time it is applied for the components that are attached to the circuit. This type of pin, as well as the roll pin, is limited to boards with a metal backing of at least 0.062″.

The third method is that of plated-through-holes. The technology available today allows the production of reliable plated-through-holes on any of the three commonly used metals presented here. Plated-through-holes were presented in Chapter 5.

Thus, we can see that the grounding of circuits for microwaves is a critical operation to perform and the metal-backed material is a viable method of obtaining this good ground contact.

With the transmission media presented and ground considerations discussed, let us proceed to describe microstrip and stripline packaging. (Recall that we will not specifically cover suspended substrate; we will refer to it, however, throughout other discussions.) Each type of packaging will be presented with pointers, suggestions, and recommendations that will aid in obtaining working microwave circuits. We should mention here that we will not go into detailed designs of packages but only indicate areas that should be watched when packaging microstrip and stripline and give suggestions to aid in the proper design of a package.

Before describing specific types of packages, we should note that every microwave package is to be treated as an individual unit. That is, treat every package as a unique design that must do a particular job. A number of requirements for microwave packages result in many nonstandard arrangements. No two nonstandard assemblies used in microwave systems are alike; each has specific requirements to perform and must fit into a certain specified space, which dictates the shape.

Note that when we speak of microwave packaging, most people think of the final outside package. This is an important part of packaging, but it is only one aspect of it. Figure 7.3 shows that there are a variety of packaging schemes that must be considered in today's integrated circuit world.

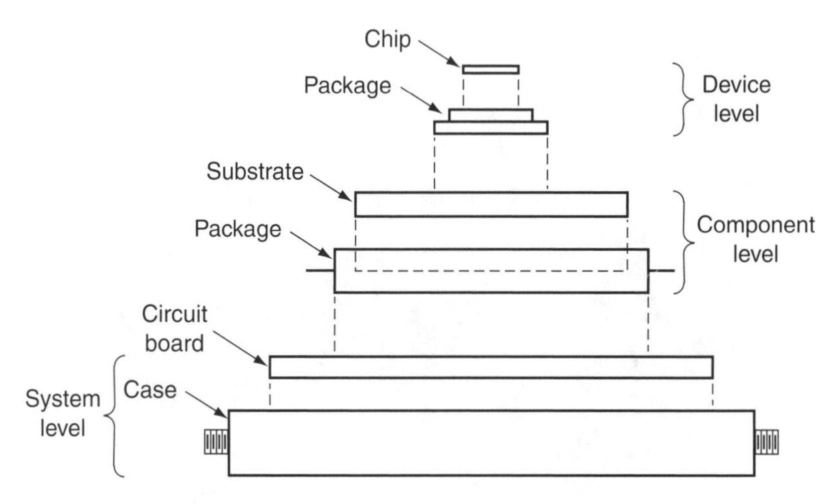

Figure 7.3 Packaging levels.

The first portion of the figure shows the actual devices that are used. These are basically the transistors. There is a chip, which often has an individual case. There are times when the chip is used by itself, but in general it has its own case. This aspect of packaging will not be presented here. It is shown to illustrate that there is a beginning level of packaging, which is often overlooked.

The next level is the packaging of components, which is probably familiar to many engineers. This level is where the circuit, with components attached to it, is placed into a case. This may be sold as is, or it may be used at a higher level of assembly. The precautions for shielding or hermeticity are often much less stringent at this level than they are at the final assembly level.

The final level can be called a system level, where a variety of modules are placed onto a main circuit board and put into the final case. This level of packaging is the most critical because it is the portion of the circuitry that is exposed to all of the external environmental factors. The system and component levels are the two areas with which we will be concerned.

The second requirement for a microwave package, other than fitting into the right space, is for it to be of very *low weight.* This is usually accomplished by building the cases from aluminum with the walls, floor, and cover as thin as practical. This aluminum is usually either silver- or gold-plated to increase conduction while maintaining the low weight (a boron-nickel plating can also be used that is lower in cost and has excellent solderability).

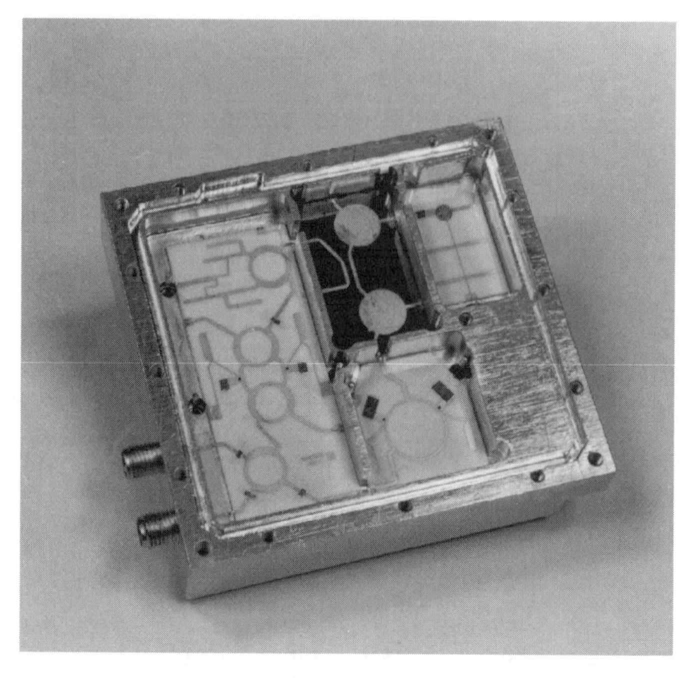

Figure 7.4 Channelized packages. Courtesy of M/A-COM.

In an effort to keep the weight of microwave packages to a minimum, and also to provide required isolation between circuits, a process called *channelizing* or *vaning* is used. Figure 7.4 shows a channelizing arrangement used to provide isolation between switch elements. You can see how such an arrangement would take a certain amount of weight from the package by having the circuit area milled from the case and at the same time providing a "tunnel" (or channel) for good isolation. Care must be taken when designing such a case, however, because improper dimensions of the channels can cause unwanted modes to be set up, which will cause problems with circuit operation. Be aware of the frequencies you are using and whether the amount of isolation required is of such a value to justify the time and expense of channelizing the case.

With some of the basic cases for microwave application presented, we can now go into more specific uses: microstrip and stripline.

7.2 Microstrip Packages

This type of packaging is probably one of the most difficult because it involves two different dielectric media and all of the rules governing each of

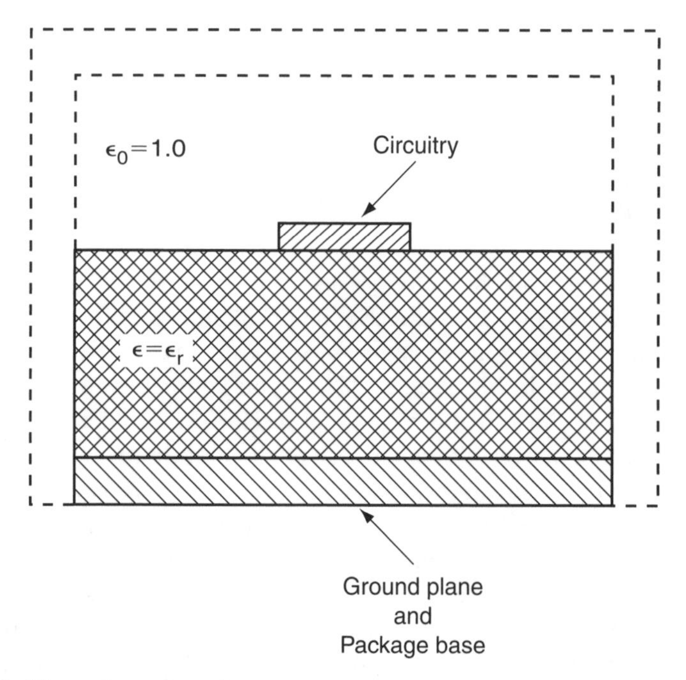

Figure 7.5 Microstrip configuration.

them. The two dielectric media are *air on top* (ε_0 = 1.0) and the *circuit dielectric* ($\varepsilon = \varepsilon_r$) itself. Figure 7.5 shows these relationships. You can see that packaging microstrip is like putting an object in an oversized box. Part of the box is filled with the object itself, and the rest is filled with air.

The first condition that must be considered when packaging microstrip circuits is the attachment of the substrate to the case. This involves much more than simply fitting a substrate to an aluminum box. The method selected will depend, to a large degree, on the material used for the substrate because the coefficient of thermal expansion becomes an important parameter when firmly attaching a substrate to a case. This parameter, of course, is the amount of movement that there is with a material during a change in temperature and is expressed as ppm/°C (parts per million per degree centigrade). You must either match coefficients of expansion of the case and substrate or provide a *buffer* material between the two if the coefficients are different.

If, for example, we are using an aluminum case and decide to use a high-dielectric ceramic-loaded PTFE material (ε_r = 10.2), we would have a choice of either mounting the material to the aluminum case by means of soldering or conductive epoxy (provided that the aluminum has been properly plated) or purchasing the material already bonded to an aluminum plate. The methods of attachment are this direct, since the coefficients of thermal expansion for high-K material ($\varepsilon_r = 10.2$) and aluminum are very similar. (Aluminum has a coefficient of expansion of 24 ppm and the high-K material is 20 to 25 ppm.) By attaching the substrate directly to the plated aluminum case, you should not encounter any problems with breaking connection either on the substrate or the connector tabs to the outside world.

If, however, you choose to use an alumina substrate in an aluminum case, the task will not be as direct. The problem arises, first of all, in the coefficients of expansion. As mentioned above, the coefficient of expansion of aluminum is 24 ppm/°C. When you compare this with alumina (130 ppm/°C), you can more fully appreciate the problem. The expansion of alumina is over five times that of the aluminum case, which suggests that there is a possibility of a solder or epoxy joint cracking if direct contact between the aluminum case and alumina substrate is made and run over any temperature range. This is the area where a buffer metal is needed, as mentioned previously in this section, and it is usually kovar. This metal was covered in Chapter 3, and you will recall that it is a combination of iron, nickel, and cobalt (all of which are ferrous metals). Its thermal properties (that is, its thermal coefficient of expansion, for example) are similar to those of alumina substrates. This metal will form a

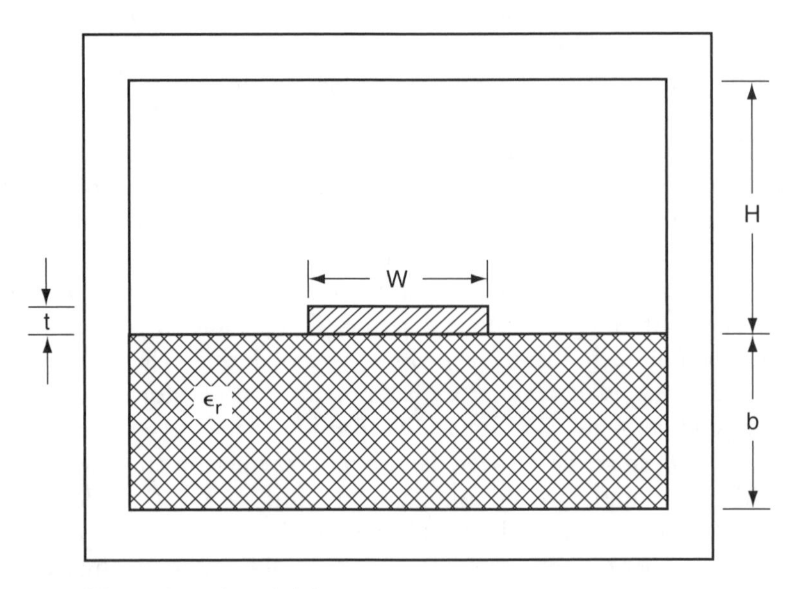

Figure 7.6 Microstrip package height.

buffer between the alumina substrate and the aluminum case to keep solder or epoxy connections from being stressed over temperature extremes, and thus is the best method for connecting materials with widely different coefficients of expansion. (In every application, you should check to be sure there are no conflicting intermetallic connections being made that will cause a questionable or unreliable connection between a substrate and case.)

A second condition that must be considered when packaging the microstrip circuits is the case height above the circuit, designated as H in Figure 7.6. This is the distance at which the top of the case is placed from the circuit; this is a critical dimension that will determine whether a final circuit will operate properly. When most microstrip circuits are breadboarded, this is usually not a primary consideration because the circuit is placed on a plate for initial testing and there is no case or top cover, as such. However, when the circuit is finally packaged, the height of the cover from the circuit is a critical dimension.

To show the importance of the H dimension, the following example is presented. Suppose we have a circuit with the following parameters:

$b = 0.025''$

$t = 0.0007''$ (½ oz copper)

$\varepsilon = 10.2$

$W = 0.060''$ (approximately a 34 to 35Ω line without the cover being considered)

To show how the impedance is effected and thus the circuit operation, we will concentrate on ε_{eff}, the effective dielectric constant. Variations in this parameter will directly affect the impedance of a microstrip line. The effective dielectric constant can be defined as:

$$\varepsilon_{eff} = \frac{\varepsilon_r + 1}{2} + q\frac{\varepsilon_r - 1}{2}$$

where:

ε_r = the relative dielectric constant of a material as given on a manufacturer's data sheet;

q = the filling factor that compensates for the two dielectric constants of microstrip, air, and the solid dielectric of the circuit.

The quantity q is further defined as:

$$q = (q_{\infty} - q_t)c_c$$

where:

q_{∞} = the filling factor for an infinite cover height;
q_t = correction for a finite conductor thickness;
q_c = correction for a noninfinite shielding.

You may have also seen the effective dielectric constant expressed in a different way:

$$\varepsilon_{eff} = 1 + q(\varepsilon_r - 1)$$

This is a valid expression and is used in microstrip design to calculate circuit parameters. The filling factor, q, in this case, does not take into account the effect of the cover (basically q_{∞}) and usually has a value that is approximately twice that of the q used in our first effective dielectric constant formula. Both equations will result in the similar effective dielectric constant values. We will show this relation later in this section. Any variations can be attributed to approximate equations and to any rounding off that may have occurred during calculations. The results should be fairly close.

With the effective dielectric constant expressed and its importance emphasized, we can proceed with the calculations of q (filling factor) that are to be used. As previously noted, there are three parts to this q: q_{∞}, q_c, and q_t. Working backwards with these terms, they can be expressed as follows:

1.) $$q_t = \frac{2 \ln 2}{\pi} \frac{t/b}{(W/b)^{1/2}}$$

2.) $$q_c = \tanh\left[1.043 + 0.121(H/b) - \frac{1.164}{(H/b)}\right]$$

3.) $$q_{\infty} = \left|1 + \frac{10b}{W}\right|^{j}$$

where:

$$j = a(W/b)\,b(\varepsilon_r)$$

and:

$$a(W/b) = 1 + \frac{1}{49}\ln\ (W/b)^2[(W/b)^2 + (1/52)^2]$$

$$[(W/b)^4 + 0.432] + \frac{1}{18.7}\ln\left[1 + \left|\frac{W}{18.1b}\right|^3\right]$$

$$b(\varepsilon_r) = -0.564\left|\frac{\varepsilon_r - 0.9}{\varepsilon_r + 3}\right|^{.053}$$

(In each of these expressions W = the strip width; t = the copper thickness; b = the dielectric thickness; H = the height of the cover above the circuit; and ε_r = the relative dielectric constant. Each is shown in Figure 7.6.)

You can see that these expressions involve several calculations, but the intent of this text is not to discuss involved mathematics. The equations are shown to illustrate what parameters (W, t, H, and b) are involved in arriving at case height: that is, q_t for the ratios of t/b and W/b; q_c for the ratio of H/b; and q_∞ for the parameters W and b with a reference made to dielectric constant (ε_r).

As mentioned, the filling factor component for case height, H, is q_c. This is the logical starting point in our example to show the effects of the cover. (For this example, we will vary H from 0.21″ to 0.36″.) Figure 7.7 shows q_c for the various values of H that we have assigned. If you refer to the expression for total q, you can see that the closer q_c is to 1.0, the less effect the cover will have on the circuit. This also can be seen in Figure 7.7.

To illustrate the full effect of the case height upon our chosen values, the parameters were put into a computer program and allowed to vary. Figures 7.8 through 7.11 show the variations in W, and thus variations in impedance, caused by varying case height and filling factor, q. The indirect effect, of course, is a variation in effective dielectric constant during this process. You will note that as the cover height, H, goes beyond the 0.26″ height, the curve does not change significantly. This value coincides closely with our rule of thumb that says we should place the cover 10 times the ground-plane spacing b away from the circuit ($10b$ = 0.250″ in our case). This chart below summarizes the results:

Height (H)	Filling Factor (q)	ε_{eff}	Ω
0.21	0.378	7.33	35.2
0.26	0.385	7.37	34.2
0.31	0.390	7.39	33.4
0.36	0.390	7.39	33.4

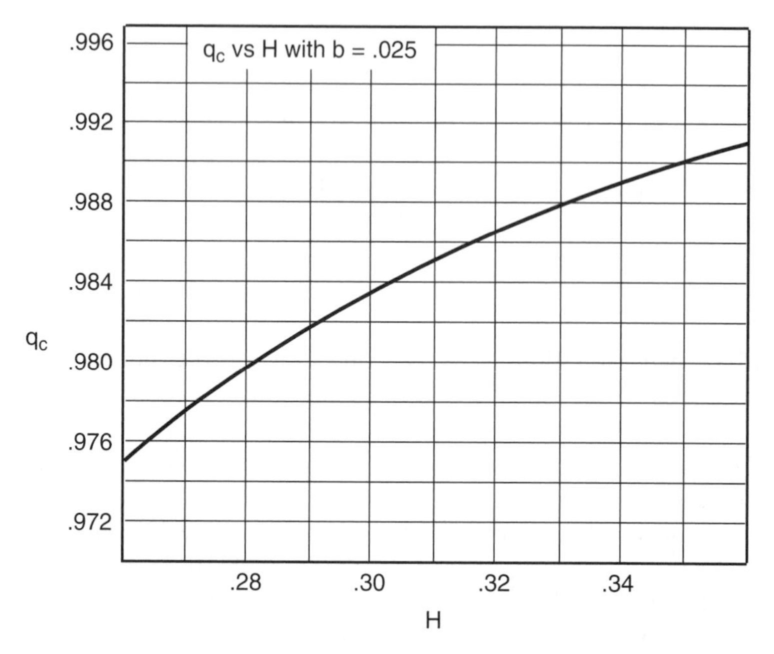

Figure 7.7 q_c versus H.

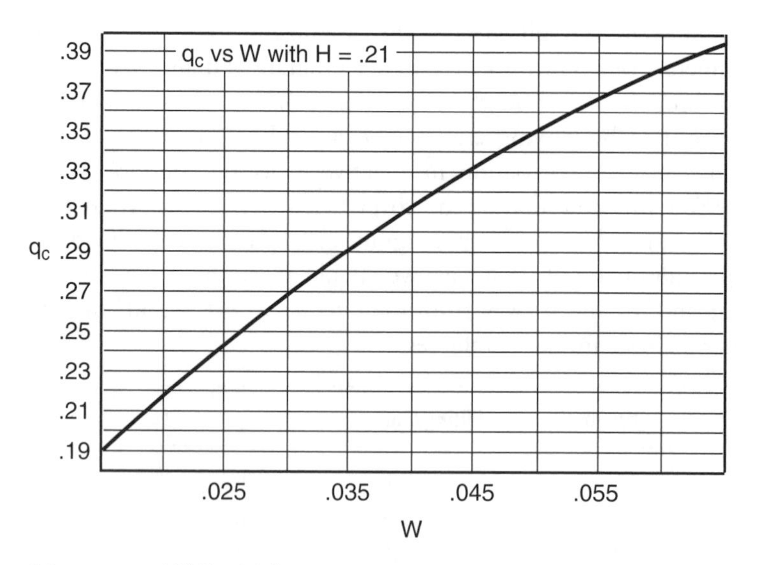

Figure 7.8 q_c versus W $(H = 0.21)$.

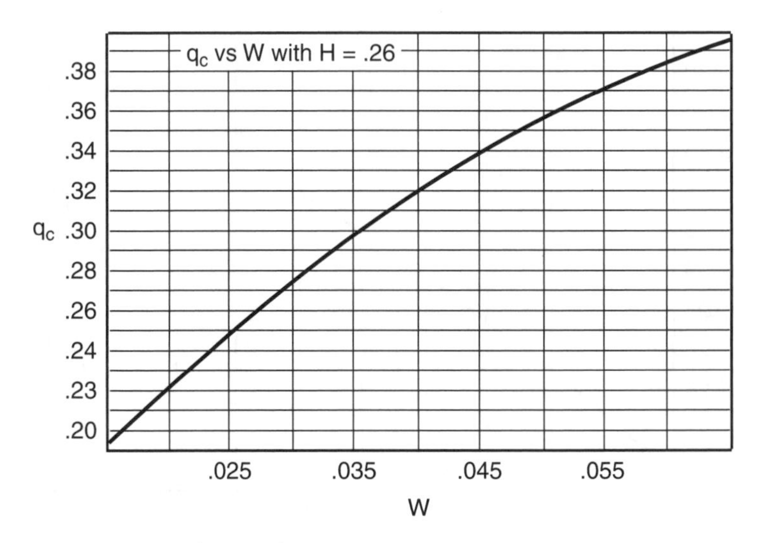

Figure 7.9 q_c versus $W(H = 0.26)$.

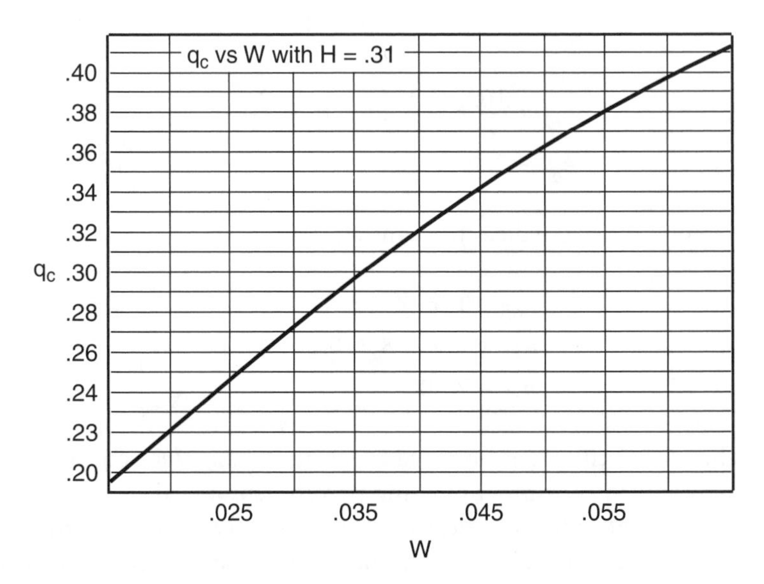

Figure 7.10 q_c versus $W(H = 0.31)$.

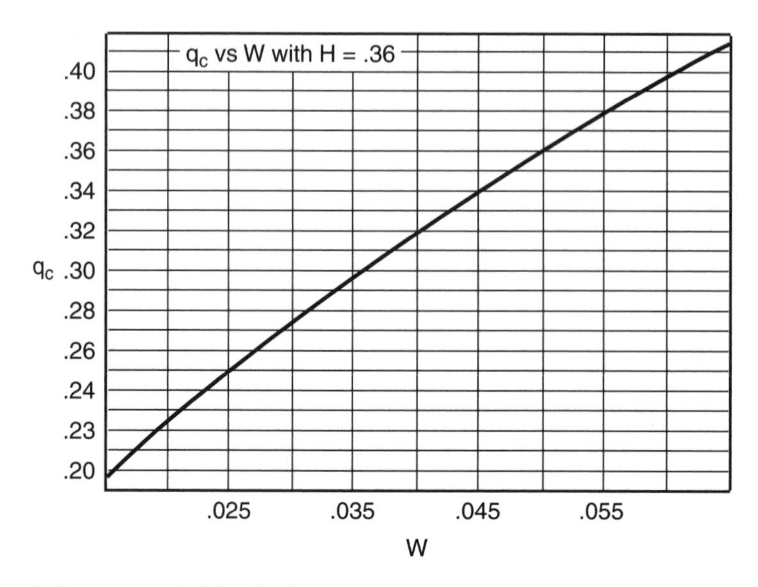

Figure 7.11 q_c versus $W(H = 0.36)$

We presented two equations for finding effective dielectric constant and said that the value for both would be similar. We now have enough information to illustrate this. The first value is:

$$\varepsilon_{eff} = \frac{\varepsilon_r + 1}{2} + q\frac{\varepsilon_r - 1}{2}$$

$$\varepsilon_{eff} = \frac{10.2 + 1}{2} + .390\,\frac{10.2 - 1}{2}$$

$$\varepsilon_{eff} = 7.39$$

For the second value, the dielectric constant is:

$$\varepsilon_{eff} = 1 + q(\varepsilon_r - 1)$$

$$\varepsilon_{eff} = 1 + .685(10.2 - 1)$$

$$\varepsilon_{eff} = 7.30$$

(The value of q used here is obtained by using Bryant-Weiss equations and obtaining a computer printout of filling factor versus impedance.)

The value of effective dielectric constant is very close for both methods. There is about a 1% difference in the two methods, which has many explanations. The point of the comparison, however, is that you can use either method and still be compensated for a cover height.

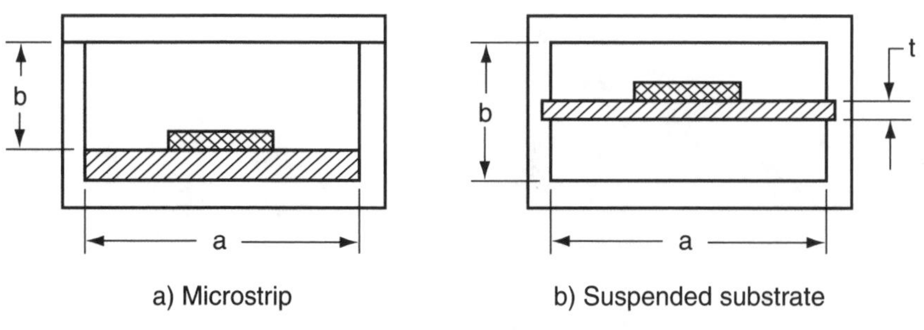

Figure 7.12 Waveguide dimensions.

A third condition that can arise in both microstrip and suspended substrate circuits is the generation of unwanted waveguide modes. As you will notice from previous figures showing microstrip and suspended substrate, they very closely resemble an end view of a piece of waveguide. For microstrip, the *a* waveguide dimension would be the horizontal distance, and the *b* dimension would be the height of the cover. For suspended substrate, the *a* and *b* dimensions would be the actual case dimensions with emphasis placed on the *t*/*b* (board thickness compared to overall case height) dimension, which will determine if a mode will be reinforced and thus cause problems. These relationships are shown in Figure 7.12. The analysis of modes and dimensions is beyond the scope of this text. If you are designing a microstrip or suspended substrate circuit, you should familiarize yourself with the calculations necessary for characterizing any unwanted modes to avoid trouble with the circuits when they are fabricated.

Some typical microstrip packaging techniques involve a variety of components integrated into one case. A milled aluminum case involves the use of a variety of ceramic substrates. In many cases, a high-K material could also be used to integrate many of the separate substrates shown in the figure. In many applications, laser welding can be used for fabrication of such a case as a microstrip assembly in a welded aluminum housing. TIG (tungsten inert gas) welding can also be used. This process allows you to weld the entire package shut without significantly raising the overall package temperature, provided that the unit is properly heat sunk.

7.3 Stripline Packaging

Stripline packages are very different from the previously covered microstrip cases. All that usually can be seen is a small flat block that has connectors on it and is painted a certain color, depending on requirements or vendor. Such small flat blocks as shown in Figure 7.13 are common packages for stripline

Figure 7.13 Stripline package. Courtesy of Anzac.

circuits. The objective of this section is to show how these packages are fitted with the microwave laminate to form a working system.

One difference between stripline and microstrip packages is that there is only one dielectric with microstrip with which to work. Just as packaging microstrip is like putting an object in an oversized box, packaging stripline is like putting an object into a custom-fit box: there is no air or excess room involved or allowed.

There are a variety of methods used for packaging stripline circuitry. We will cover four basic types of packages and make reference to any variations that may be used: the *sandwich package* used for breadboarding, *channelized chassis, box and cover,* and the popular *caseless package.*

The first type of package is shown in Figure 7.14 in a side view. A close look at the construction will show why it is excellent for stripline breadboarding. The package consists of an aluminum plate on top, the two dielectric pieces (or three if a thin dielectric is used for the circuit as in the

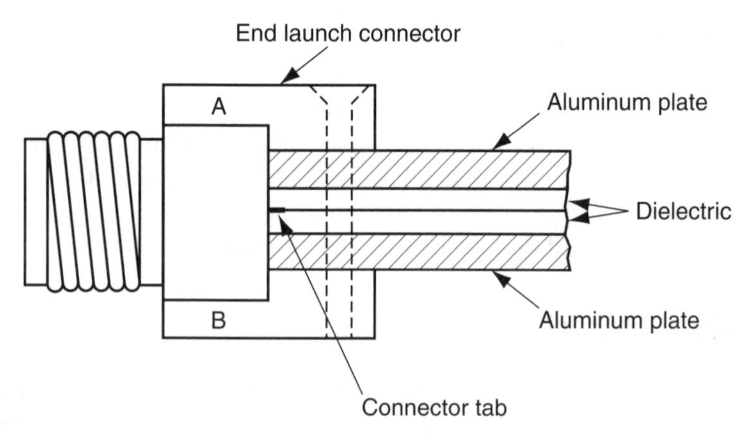

Figure 7.14 Sandwich-style package.

case of a quadrature hybrid), and the bottom aluminum plate. The construction fulfills all of the criteria necessary for stripline operation.

The connector is different from the typical package connectors used for the type of enclosure usually associated with stripline and microwave circuits. The usual connector may have either a two- or four-hole flange that is attached to the metallic portion of the case with the center conductor attached to the circuit. In another arrangement, the only part of the connector is the threaded portion that is inserted into the case, which is tapped to accept the connector and have the center conductor, once again, attached to the circuit. The connector in Figure 7.14 does neither of these things; instead it forms a clamp to hold the stripline package together and apply very tight pressure on the center conductor to hold it on the circuit. This type of connector is called an *end launch* connector. Spacings available between the top plate of the connector (A in Figure 7.14) and the bottom plate (B) are 0.062, 0.125, and 0.250″. Those most widely used are the 0.125 and 0.250″. These allow you to use two pieces of 0.030″ PTFE/glass laminates with two pieces of 0.030″ aluminum in one case, and two pieces of 0.062″ PTFE/glass laminate with two pieces of 0.062″ aluminum in the other. This connector will allow you to assemble a package for breadboard and easily disassemble it to make changes.

A variation of the stripline package covered above is called *flat plate* construction. In this type of construction, a flange-mounted connector is used because the plates wrap around the circuit and allow this type of connector to be used. It is a bit more difficult to assemble and disassemble but still has an advantage of simplicity.

A second type of stripline package is the channelized chassis. A case has been milled with channels that accept the stripline laminates. The circuits are flush with the case and are not set on top of it. These cases can either be machined or stamped to the needed configuration. The paths into which the circuit will be placed are milled to preserve the wall proximity on both sides of the circuit. This type of construction is very complex and can be expensive. Therefore, it is generally used only where the circuit performance necessitates and justifies the high cost.

A third method, called box and cover, is a popular type of stripline package. The circuits shown in Figure 7.15 are all fabricated by using this type of construction. The method is similar to the channelized chassis in that the board is cut. In the box and cover chassis, however, the board is not cut to conform to the circuit pattern but rather to the outside edge (border) of the board. In this way, the circuit is dropped into the case as before, but this method does not involve the complexity of cutting, as in the case of the

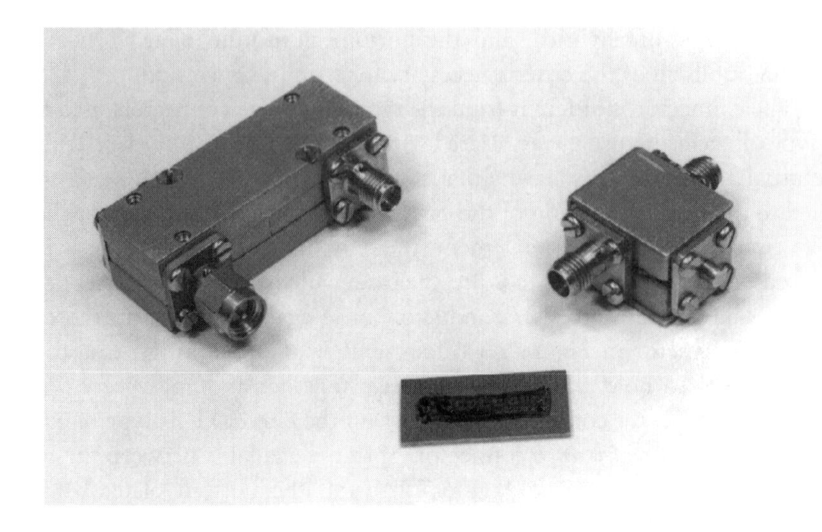

Figure 7.15 Stripline package. Courtesy of Anaren Microwave.

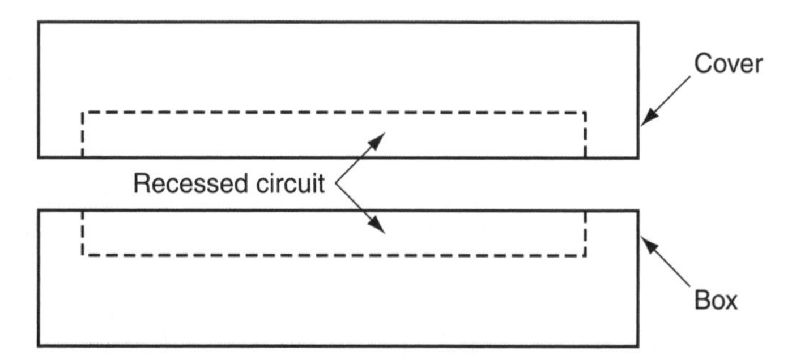

Figure 7.16 Box and cover chassis.

channelized chassis. This is shown in Figure 7.16. You can see that the "box" and "cover" are each machined so that the circuit board (laminate) can be recessed into the metal itself, resulting in a firm, tight fit of the stripline package, which probably explains why it is so widely used throughout the microwave industry where stripline is fabricated.

Figure 7.17 shows how connectors can be put into the box and cover type of case. Generally, the barrel of the connector is extended from the conventional type of connector so that it can be inserted into the case without having threads inside the metal portion of the case. A slot (shown in the figure) is cut in the channel to accommodate the flange portion of the connector (the flange is equivalent to the two-hole flange size). This machining

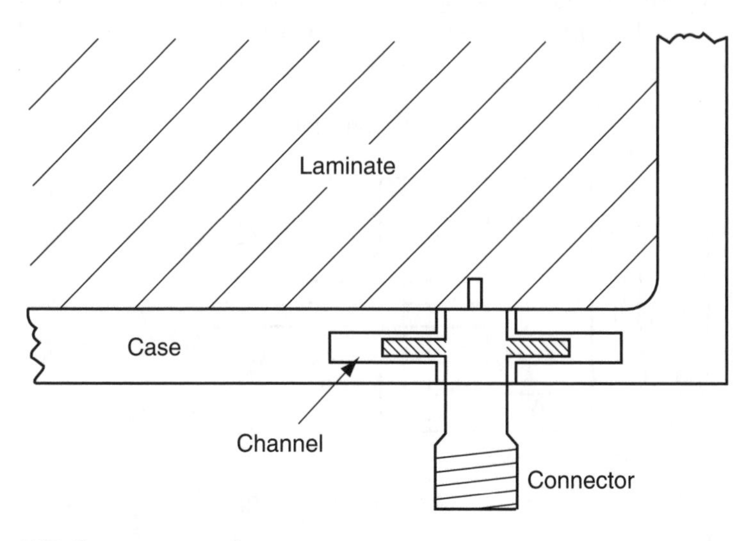

Figure 7.17 Connector mounting.

arrangement gives the connector a solid place to set itself and thus eliminates any twist (or torque) on the connector that could break the connector tab away from the circuit.

The final type of stripline package to be covered is called the *no chassis* or "caseless" package. This type of package is useful when drop-in components are used. Figure 7.18 shows such a case with a connector attached. The case may have connectors (as shown) or simply tabs coming from the ports.

The package is assembled by placing the ground-plane laminates together with the thin dielectric material between them. Aluminum plates are placed on the top and bottom of the ground-plane laminates, and heat and pressure are applied (thermocompression process) until the package is fused together. Sometimes an adhesive or a bonding film is placed between layers prior to the thermocompression phase.

The obvious advantage of such a package is that no complex machining is necessary. One large disadvantage is that the unit is not reparable. If a problem arises, the unit must be scrapped and replaced with a new one. You will completely tear the laminates apart before you even begin to separate them, if you are so inclined to try. Do not use this package for active devices unless you intend to make it a throw-away module. Also, be sure that your active devices will take the heat involved in the assembly of such a case; most active devices will not take the lamination temperatures. Thus, the caseless type chassis seems to be limited to passive devices.

There are a variety of case styles that can be used for stripline circuits, depending on your particular application. Take great care in choosing the style you use because it can either make or break your carefully built mi-

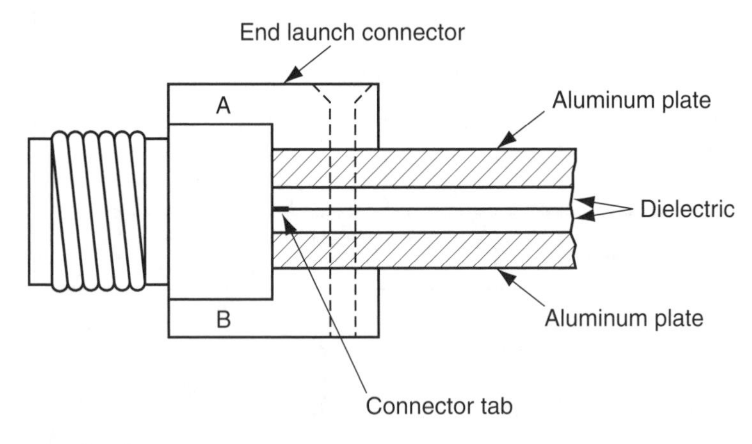

Figure 7.18 Caseless package.

Table 7.1
Comparison of Hybrid and MMIC Technologies

	Hybrids	MMICs
Advantages	Low-cost substrates	Small size and weight
	Lower capital costs	Better reliability and reproducibility
	Easily repaired	Broader frequency performance
	Easier testing and tuning	Low cost in large quantities
	Lower losses	Circuit design flexibility
	Higher q (filling factor)	
Disadvantages	Lower reliability	Higher losses
	Limited bandwidth	Circuit tuning not practical
	Uncontrolled parasitics	Undesired RF coupling
	Limited frequency range	Poor heat sinking
	Larger size	High parasitic capacitance
	More piece parts	High cost
	High assembly cost	Young processing technology
		High capital equipment cost

crowave design. There are times when the medium used for transmission of microwave energy is not the stripline or microstrip configuration (hybrid), but the MMIC (monolithic microwave integrated circuit). There are advantages and disadvantages to each method, hybrid or MMIC technology, and these characteristics are shown in Table 7.1.

With these advantages and disadvantages, one can see that there are definite trade-offs when considering which medium to use. The use of MMIC technology incorporates many of the ideas put forth for the microstrip packaging. We still need a good ground connection, we still must be concerned about the height of the cover above the circuit because the MMIC circuitry is a microstrip configuration, and connections to other circuits and the outside world remain critical.

7.4 Package Sealing

The packages covered in this chapter for stripline, microstrip, and suspended substrate generally have to protect the internal circuit from elements such as moisture, humidity, salt spray, dust, and outside electrical interference (electromagnetic interference, EMI). In order to protect the circuit adequately, certain methods of *sealing* can be used. Some of these methods are presented here:

- *Conductive epoxy*—This is an excellent sealer that can be put around the outside edge of the case and cured to make the box essentially one piece. This method of sealing generally makes the assembly irreparable.
- *Solder*—This is a very effective method of sealing. Low-temperature solder should be used so that repair can be made on the package if necessary.
- *Gasket material*—This can be placed in the case as it is put together to form an excellent seal. Care should be taken to ensure the material is not too thick, which could upset the ground-plane spacing of the stripline circuit.
- *Welding*—This is becoming more popular as a means of sealing microwave packages. It is very effective and also allows for a certain number of repairs.
- *Metallized tape*—This is an inexpensive method of sealing a microwave package. Commercial applications will use such a method. It is available in both conductive and nonconductive adhesive and can also utilize thermosetting adhesives that can be baked, resulting in even better sealing.

7.5 Microwave Connectors

An important part of any microwave packaging assignment is to choose the proper connector for that particular component or system.

The microwave connector should be treated as a separate microwave component and, as such, should be given the same extensive consideration required when choosing a directional coupler, for example, for a specific test setup. Never underestimate the importance of connectors to the operation of your circuit. Many times, a circuit is assumed to have a problem when a bad connector on the case is keeping the circuit from operating properly. In this case, simple connector replacement will cause all systems to operate properly to their expected level of performance and allow you to go on to new and exciting challenges.

Appendix G has five common types of connectors that are all 50 ohm connectors and are used in microwave applications. They are the SMA, TNC, Type N, APC-7, and the K connectors. The SMA is probably the most common type of microwave connector and the one most often seen on microwave packages. It is a small connector that performs microwave functions very well.

The TNC connector is actually a version of the common BNC connector. The main difference is that it has threads on it to allow the connector to have a much more efficient grounding system at higher microwave frequencies.

The Type N connector is a larger microwave connector that is used many times for some of the lower microwave frequency bands. The APC-7 is a sexless connector that has the two sections of the connector butting up to one another and making an excellent connection for a wide range of frequencies.

The K connector is a much higher frequency connector. Many times, this connector is used beyond the microwave band (20 GHz) and proceeds upwards to the 40 GHz region. This extended range makes the K connector an excellent microwave connector also.

7.6 EMI Considerations

When considering packaging, it is necessary to not only put the circuit in a case that will provide sufficient grounding and adequate input/output connections but also necessary (and critical) to provide a case that will keep the signals in the circuit where they belong and keep all other signals out. This is where electromagnetic interference (EMI) considerations come into play. EMI can be very harmful to an electronic circuit. In the same vein, electrostatic discharge (ESD) is also something that you do not want your microwave circuit exposed to.

EMI is generated wherever there are rapidly changing electric and magnetic fields. Common sources are fluorescent lights, electric motors, and common household appliances. These are all man-made sources of EMI. There also is a very large EMI source—lightning discharges—that is not man-made. Many pieces of electronic communications equipment have been damaged by lightning discharges. (Lightning is also many times considered to be a source of ESD.)

ESD is generated by friction between two surfaces that cause a voltage difference to be produced between them. These charges can reach up into the kilovolt range under dry conditions. This can be looked at as a miniature lightning strike that has some of the same devastating effects as a full-blown lightning strike. In many cases, ESD is even more destructive since it operates on some of the very thin device layers of surface mount devices (SMD) that are so popular today. This thin layer is easily penetrated and destroyed by these charges.

Knowing about EMI or ESD is one thing. Knowing how to reduce, or eliminate, them is an entirely different problem. Just how do you reduce this phenomena that causes so much damage to sensitive microwave components? Certain guidelines should be observed when putting together a circuit and case design that will reduce the effects of these destructive elements. A few of these guidelines are shown here:

1. Use a multilayer circuit board that has separate grounds for the microwave signal and the voltages used.
2. Use feedthrough capacitors at the entry and exit points.
3. Terminate striplines and microstrip lines with the proper impedance to reduce reflections.
4. Keep all component leads short or use components that have terminations on them rather than leads.
5. Be sure that all cables have a full ground on them to prevent any radiation of signals at the ungrounded areas.
6. Use a series resistor to limit the current to any stage that is a high impedance stage. This will limit the current.
7. Use a small ceramic capacitor to decouple sensitive points in the circuit to ground. Mount them close to the leads or connection points.
8. Group components on a circuit board according to their frequency of operation.
9. Isolate analog and digital circuits.
10. Place appropriate ground between the signal lines and any possible EMI signal generating lines.
11. Provide large ground plane areas on the circuit board.
12. Keep high power circuits away from low power circuits.
13. Provide a minimum spacing between traces on a microwave circuit board. The general rule of thumb is to have lines separated by at least one ground plane spacing.

Just about everything we stated about EMI and ESD to this point has involved the circuit board and some of the ways problems could be reduced. There are, however, many cases where it is not possible to solve EMI problems at the circuit board level. It becomes necessary to provide shielding at the case level—that is, cases, connectors, and cables.

One area overlooked many times when dealing with EMI requirements is openings or doors that may be on a cabinet for a system. The doors and panels of these units need to be connected electrically on all sides. This should be done with EMI shielding gaskets. Many of these gaskets are self-adhesive and easy to install. The gasket material that is used should have many aspects taken into consideration. Such conditions as the rigidity of the construction, the distance between the boltings (where the bolts go through the chassis to hold the shielding), and the distance between the bolting area and the construction materials used should be considered.

Another area susceptible to EMI is the area where displays and vent panels are installed. Displays can be purchased with a sputtered transparent conductive coating for shielding. Vent panels can be shielded with aluminum honeycomb vents that provide excellent shielding and yet do not obstruct the air flow to the equipment.

Cables are a source of problems for EMI radiation. Cables should be shielded with either ready-made shielding, shielding tubes that can be put over the cable, or a form of cable wrapping. When you shield cables, you must complete the job by connecting the shielding properly to the shielding of the enclosure itself.

Connectors should also be shielded or filtered in order to reduce EMI. This can be a requirement when actually specifying which connector to use for your particular application. Also, any unused connector should be capped in order to prevent any stray signals from getting into that connector and, eventually, to other parts of the circuit.

7.7 Chapter Summary

This chapter covered a very important step in the fabrication of microwave devices and circuits: packages. Microwave packages are much more than just a box in which to put the circuit: they are an integral part of the overall circuit design. When your ultimate circuit design is completed, take at least as much time working on the package as you did on the circuit. It will pay off handsomely in the long run.

The chapter ended up with a section on EMI and ESD. These are real problem areas in many microwave circuits and should be addressed thoroughly.

Appendix A
Dielectric Constants

Material	Dielectric Constant
Air	1.00
Teflon (PTFE)	2.10
Polyethylene	2.25
Rexolite No. 1422	2.55
Neoprene Rubber	4.00
Balsawood	1.30
Snow (freshly fallen)	1.20
Snow (hard packed)	1.50
Water (distilled)	76.70

Appendix B
Coefficients of Expansion

Element	Expansion (ppm)
Aluminum	24.3
Carbon	0.6–4.3
Chromium	6.2
Cobalt	12.3
Copper	16.56
Gold	14.2
Indium	33.0
Iron	11.7
Lead	28.7
Magnesium	25.2
Nickel	13.3
Silver	18.9
Tin	23.0
Titanium	8.5

ppm = parts per million

Appendix C
Chemical Symbols

Element	Symbol
Aluminum	Al
Beryllium	Be
Boron	B
Carbon	C
Chromium	Cr
Cobalt	Co
Copper	Cu
Gallium	Ga
Germanium	Ge
Gold	Au
Indium	In
Iron	Fe
Lead	Pb
Magnesium	Mg
Nickel	Ni
Platinum	Pt
Silicon	Si
Silver	Ag
Tin	Sn
Titanium	Ti
Zinc	Zn

Appendix D
Melting Points

Material	Melting Point (°C)
Aluminum	660
Cobalt	1495
Copper	1083
Gold	1063
Indium	156
Iron	1535
Lead	327
Nickel	1453
Silver	960
Tin	232

Appendix E
Temperature Conversion

Table AE.1

−459.4 to 0			0 to 100					
C		F	C		F	C		F
−273	−459.4		−17.8	0	32	10.0	50	122.0
−268	−450		−17.2	1	33.8	10.6	51	123.8
−262	−400		−16.7	2	35.6	11.1	52	125.6
−257	−430		−16.1	3	37.4	11.7	53	127.4
−251	−420		−15.6	4	39.2	12.2	54	129.2
−246	−410		−15.0	5	41.0	12.8	55	131.0
−240	−400		−14.4	6	42.8	13.3	56	132.8
−234	−390		−13.9	7	44.6	13.9	57	134.6
−229	−380		−13.3	8	46.4	14.4	58	136.4
−223	−370		−12.8	9	48.2	15.0	59	138.2
−218	−360		−12.2	10	50.0	15.6	60	140.0
−212	−350		−11.7	11	51.8	16.1	61	141.8
−207	−340		−11.1	12	53.6	16.7	62	143.6
−201	−330		−10.6	13	55.4	17.2	63	145.4
−196	−320		−10.0	14	57.2	17.8	64	147.2
−190	−310		− 9.4	15	59.0	18.3	65	149.0
−184	−300		− 8.9	16	60.8	18.9	66	150.8
−179	−290		− 8.3	17	62.6	19.4	67	152.6
−173	−280		− 7.8	18	64.4	20.0	68	154.4
−169	−273	−459.4	− 7.2	19	66.2	20.6	69	156.2
−168	−270	−454	− 6.7	20	68.0	21.1	70	158.0
−162	−260	−436	− 6.1	21	69.8	21.7	71	159.8

Table AE.1 (continued)

−459.4 to 0			0 to 100					
C		F	C		F	C		F
−157	−250	−418	− 5.6	22	71.6	22.2	72	161.6
−151	−240	−400	− 5.0	23	73.4	22.8	73	163.4
−146	−230	−382	− 4.4	24	75.2	23.3	74	165.2
−140	−220	−364	− 3.9	25	77.0	23.9	75	167.0
−134	−210	−346	− 3.3	26	78.8	24.4	76	168.8
−129	−200	−328	− 2.8	27	80.6	25.0	77	170.6
−123	−190	−310	− 2.2	28	82.4	25.6	78	172.4
−118	−180	−292	− 1.7	29	84.2	26.1	79	174.2
−112	−170	−274	− 1.1	30	86.0	26.7	80	176.0
−107	−160	−256	− 0.6	31	87.8	27.2	81	177.8
−101	−150	−238	0.0	32	89.6	27.8	82	179.6
− 96	−140	−220	0.6	33	91.4	28.3	83	181.4
− 90	−130	−202	1.1	34	93.2	28.9	84	183.2
− 84	−120	−184	1.7	35	95.0	29.4	85	185.0
− 79	−110	−166	2.2	36	96.8	30.0	86	186.8
− 73	−100	−148	2.8	37	98.6	30.6	87	188.6
− 68	− 90	−130	3.3	38	100.4	31.1	88	190.4
− 62	− 80	−112	3.9	39	102.2	31.7	89	192.2
− 57	− 70	− 94	4.4	40	104.0	32.2	90	194.0
− 51	− 60	− 65	5.0	41	105.8	32.8	91	195.8
− 46	− 50	− 58	5.6	42	107.6	33.3	92	197.6
− 40	− 40	− 40	6.1	43	109.4	33.9	93	199.4
− 34	− 30	− 22	6.7	44	111.2	34.4	94	201.2
− 29	− 20	− 4	7.2	45	113.0	35.0	95	203.0
− 23	− 10	+ 14	7.8	46	114.8	35.6	96	204.8
− 17.8	− 0	+ 32	8.3	47	116.6	36.1	97	206.6
			8.9	48	118.4	36.7	98	208.4
			9.4	49	120.2	37.2	99	210.2
						37.8	100	212.0

Table AE.2

100 to 1000						1000 to 2000					
C		F	C		F	C		F	C		F
38	100	212	260	500	932	538	1000	1832	816	1500	2732
43	110	230	266	510	950	543	1010	1850	821	1510	2750
49	120	248	271	520	968	549	1020	1868	827	1520	2768
54	130	266	277	530	986	554	1030	1886	832	1530	2786
60	140	284	282	540	1004	560	1040	1904	838	1540	2804
66	150	302	288	550	1022	566	1050	1922	843	1550	2822
71	160	320	293	560	1040	571	1060	1940	849	1560	2840
77	170	338	299	570	1058	577	1070	1958	854	1570	2858
82	180	356	304	580	1076	582	1080	1976	860	1580	2876
88	190	374	310	590	1094	588	1090	1994	866	1590	2894
93	200	392	316	600	1112	593	1100	2012	871	1600	2912
99	210	410	321	610	1130	599	1110	2030	877	1610	2930
100	212	413.6	327	620	1148	604	1120	2048	882	1620	2948
104	220	428	332	630	1166	610	1130	2066	888	1630	2966
110	230	446	338	640	1184	616	1140	2084	893	1640	2984
116	240	464	343	650	1202	621	1150	2102	899	1650	3002
121	250	482	349	660	1220	627	1160	2120	904	1660	3020
127	260	500	354	670	1238	632	1170	2138	910	1670	3038
132	270	518	360	680	1256	638	1180	2156	916	1680	3056
138	280	536	366	690	1274	643	1190	2174	921	1690	3074
143	290	554	371	700	1292	649	1200	2192	927	1700	3092
149	300	572	377	710	1310	654	1210	2210	932	1710	3110
154	310	590	382	720	1328	660	1220	2228	938	1720	3128
160	320	608	388	730	1346	666	1230	2246	943	1730	3146
166	330	626	393	740	1364	671	1240	2264	949	1740	3164
171	340	644	399	750	1382	677	1250	2282	954	1750	3182
177	350	662	404	760	1400	682	1260	2300	960	1760	3200
182	360	680	410	770	1418	688	1270	2318	966	1770	3218
188	370	698	416	780	1436	693	1280	2336	971	1780	3236
193	380	716	421	790	1454	699	1290	2354	977	1790	3254
199	390	734	427	800	1472	704	1300	2372	982	1800	3272
204	400	752	432	810	1490	710	1310	2390	988	1810	3290

Table AE.2 (continued)

100 to 1000						1000 to 2000					
C		F	C		F	C		F	C		F
210	410	770	438	820	1508	716	1320	2408	993	1820	3308
216	420	788	443	830	1526	721	1330	2426	999	1830	3326
221	430	806	449	840	1544	727	1340	2444	1004	1840	3344
227	440	824	454	850	1562	732	1350	2462	1010	1850	3362
232	450	842	460	860	1580	738	1360	2480	1016	1860	3380
238	460	860	466	870	1598	743	1370	2498	1021	1870	3398
243	470	878	471	880	1616	749	1380	2516	1027	1880	3416
249	480	896	477	890	1634	754	1390	2534	1032	1890	3434
254	490	914	482	900	1652	760	1400	2552	1038	1900	3452
			488	910	1670	766	1410	2570	1043	1910	3470
			493	920	1688	771	1420	2588	1049	1920	3488
			499	930	1706	777	1430	2606	1054	1930	3506
			504	940	1724	782	1440	2624	1060	1940	3524
			510	950	1742	788	1450	2642	1066	1950	3542
			516	960	1760	793	1460	2660	1071	1960	3560
			521	970	1778	799	1470	2678	1077	1970	3578
			527	980	1796	804	1480	2696	1082	1980	3596
			532	990	1814	810	1490	2714	1088	1990	3614
			538	1000	1832				1093	2000	3632

Appendix F
Data Sheets For Microwave Laminates

Microwave Materials

PTFE/Woven Fiberglass Laminates: Microwave Printed Circuit Board Substrates

AR320™ is a woven fiberglass, reinforced PTFE based composite material for use as a printed circuit board substrate. It is available in thicknesses of 0.024" and greater.

The higher weight ratio of fiberglass to PTFE yields a laminate with greater dimensional stability than is normally expected with PTFE-based substrates. Additionally, the use of heavier grades of fiberglass and the displacement of a significant portion of PTFE reduces the overall cost of the laminate. This affords "PTFE performance" at a more economical cost.

The moderately higher dielectric constant of AR320 also allows a marginal reduction in circuit size for a typical microwave circuit. The low loss makes AR320 ideal for digital signal processing applications.

AR320 is also consistent with the processing used for standard PTFE based printed circuit board substrates. Because there is a relatively higher amount of fiberglass, thermal expansion is reduced in all directions. This improves plated through hole reliability.

Availability:

AR320 laminates are supplied with 1/2, 1 or 2 ounce electrodeposited copper on both sides. Other copper weights and rolled copper foil are available. AR320 is available bonded to a heavy metal ground plane. Aluminum, brass or copper plates also provide an integral heat sink and mechanical support to the substrate.

When ordering AR320 products, please specify thickness, cladding, panel size and any other special considerations. Available master sheet sizes include 36" x 48", 36" x 72" and 48" x 54".

MATERIALS FOR ELECTRONICS

Typical Properties: AR320™ PTFE/Woven Fiberglass Laminates

Properties	Test Method	Condition	Typical Values
Dielectric Constant @10GHz	IPC TM-650 2.5.5.5	C23/50	3.20
Dissipation Factor @10GHz	IPC TM-650 2.5.5.5	C23/50	0.003
Thermal Coefficient o E_r (ppm/°C)	IPC TM-650 2.5.5.5 Adapted	-10°C to +140°C	-100
Volume Resistivity (MΩ-cm)	IPC TM-650 2.5.17.1	C96/35/90	1.2×10^9
Surface Resistivity (MΩ)	IPC TM-650 2.5.17.1	C96/35/90	4.5×10^7
Arc Resistance (seconds)	ASTM D-495	D48/50	> 180
Tensile Modulus (kpsi)	ASTM D-638	A, 23°C	706, 517
Tensile Strength (kpsi)	ASTM D-882	A, 23°C	23.6, 21.5
Compressive Modulus (kpsi)	ASTM D-695	A, 23°C	372
Flexural Modulus (kpsi)	ASTM D-790	A, 23°C	545
Dielectric Breakdown (kv)	ASTM D-149	D48/50	> 45
Specific Gravity (g/cm^3)	ASTM D-792 Method A	A, 23°C	2.45
Water Absorption (%)	MIL-S-13949H 3.7.7 IPC TM-650 2.6.2.2	E1/105 + D24/23	0.08
Coefficient of Thermal Expansion (ppm/°C) X Axis Y Axis Z Axis	IPC TM-650 2.4.24 Mettler 3000 Thermomechanical Analyzer	0°C to 100°C	 9 12 71
Thermal Conductivity (W/mK)	ASTM E-1225	100°C	0.230
Outgassing Total Mass Loss (%) Collected Volatile Condensable Material (%) Water Vapor Regain (%) Visible Condensate (±)	NASA SP-R-0022A Maximum 1.00% Maximum 0.10%	 125°C, ≤ 10^{-6} torr	 0.02 0.00 0.01 NO
Flammability (UL File E 80166)	UL 94 Vertical Burn IPC TM-650 2.3.10	C48/23/50, E24/125	UL94V-0

Data based on 0.062" dielectric thickness, exclusive of metal cladding except where indicated by test method. Results listed above are typical properties; they are not to be used as specification limits. The above information creates no expressed or implied warranties. The properties of AR320 laminates may vary depending on the application.

MATERIALS FOR ELECTRONICS

1100 Governor Lea Road, Bear, DE 19701 • Telephone: (302) 834-2100, (800) 635-9333 • Fax: (302) 834-2574
9433 Hyssop Drive, Rancho Cucamonga, CA 91730 • Telephone: (909) 987-9533 • Fax: (909) 987-8541
37 Rue Collange, 92300 LeVallois, Perret, France • Telephone: (33) 1-427-02642 • Fax: (33) 1-427-02798
44 Wilby Avenue, Little Lever, Bolton, Lancashire, BL31QE, U.K. • Telephone: (44) 120-457-6068 • Fax: (44) 120-479-6463
E-mail: substrates@arlonmed.com • Website: www.arlonmed.com

Arlon is an ISO 9002 Registered Company

AR350™/AR450™

Microwave Materials

PTFE/Nonwoven Fiberglass/Ceramic Laminates: Circuit Board Substrates for High Speed and Microwave Applications

AR350™ and AR450™ represent a class of PTFE based laminates designed to offer dielectric constants similar to thermoset laminates. These materials offer nominal dielectric constants of 3.50 and 4.50, approximating the values of such resin systems as BT, cyanate ester and FR-4 epoxy. Since the dielectric constants match the existing thermoset materials, designs on these products would not require modification to take advantage of the lower loss properties available with the AR350 and AR450 products.

The nominal dielectric constant tolerance of ± 0.03 is more consistent than that of typical thermoset grades, assuring more consistent performance. Another significant advantage offered by these products includes the loss tangent performance a designer would expect with PTFE based laminates, up to an order of magnitude better than thermoset resin based materials.

Adapting coating technology from PTFE/glass laminates, AR350 and AR450 have much better dielectric constant uniformity, both within a sheet and between sheets, than thermoset based laminates of similar dielectric constants. To the designer, dielectric constant uniformity and low loss tangent allow better impedance control and a higher signal to noise ratio.

Availability:

AR350 and AR450 laminates are supplied with 1/2, 1 or 2 ounce electrodeposited copper on both sides. Other copper weights and rolled copper foil are available. AR350 and AR450 are available bonded to a heavy metal ground plane. Aluminum, brass or copper plates also provide an integral heat sink and mechanical support to the substrate.

When ordering AR350 and AR450 products, please specify thickness, cladding, panel size and any other special considerations. Available master sheet sizes include 36" x 48" and 36" x 72".

Typical Properties: AR350™ and AR450™ PTFE/Nonwoven Fiberglass/Ceramic Laminates

Properties	Test Method	Condition	Typical Values AR350	Typical Values AR450
Dielectric Constant @10GHz	IPC TM-650 2.5.5.5	C23/50	3.50	4.50
Dissipation Factor @10GHz	IPC TM-650 2.5.5.5	C23/50	0.0026	0.0035
Thermal Coefficient of E_r (ppm/°C)	IPC TM-650 2.5.5.5 Adapted	-10°C to +140°C	-213	-238
Volume Resistivity (MΩ-cm)	IPC TM-650 2.5.17.1	C96/35/90	3.4×10^{13}	3.3×10^{13}
Surface Resistivity (MΩ)	IPC TM-650 2.5.17.1	C96/35/90	4.6×10^{11}	4.8×10^{11}
Arc Resistance (seconds)	ASTM D-495	D48/50	> 180	> 180
Tensile Modulus (kpsi)	ASTM D-638	A, 23°C	154, 147	155, 150
Tensile Strength (kpsi)	ASTM D-882	A, 23°C	5.1, 3.9	2.2, 2.0
Compressive Modulus (kpsi)	ASTM D-695	A, 23°C	223	228
Flexural Modulus (kpsi)	ASTM D-790	A, 23°C	342	345
Dielectric Breakdown (kv)	ASTM D-149	D48/50	> 45	> 45
Specific Gravity (g/cm^3)	ASTM D-792 Method A	A, 23°C	2.36	2.39
Water Absorption (%)	MIL-S-13949H 3.7.7 IPC TM-650 2.6.2.2	E1/105 + D24/23	0.08	0.08
Coefficient of Thermal Expansion (ppm/°C)	IPC TM-650 2.4.24 Mettler 3000 Thermomechanical Analyzer	0°C to 100°C		
X Axis			35	30
Y Axis			35	30
Z Axis			107	102
Thermal Conductivity (W/mK)	ASTM E-1225	100°C	0.310	0.320
Outgassing	NASA SP-R-0022A			
Total Mass Loss (%)	Maximum 1.00%	125°C, ≤ 10^{-6} torr	0.04	0.04
Collected Volatile Condensable Material (%)	Maximum 0.10%		0.01	0.01
Water Vapor Regain (%)			0.03	0.03
Visible Condensate (±)			NO	NO
Flammability	UL 94 Vertical Burn IPC TM-650 2.3.10	C48/23/50, E24/125	Meets requirements of UL94V-0	Meets requirements of UL94V-0

Data based on 0.062" dielectric thickness, exclusive of metal cladding except where indicated by test method. Results listed above are typical properties; they are not to be used as specification limits. The above information creates no expressed or implied warranties. The properties of AR350 and AR450 laminates may vary depending on the application.

The information and data contained herein are believed reliable, but all recommendations or suggestions are made without guarantee. You should thoroughly and independently test materials for any planned applications and determine satisfactory performance before commercialization. Furthermore, no suggestion for use, or material supplied shall be construed as a recommendation or inducement to violate any law or infringe any patent.

MATERIALS FOR ELECTRONICS

1100 Governor Lea Road, Bear, DE 19701 • Telephone: (302) 834-2100, (800) 635-9333 • Fax: (302) 834-2574
9433 Hyssop Drive, Rancho Cucamonga, CA 91730 • Telephone: (909) 987-9533 • Fax: (909) 987-8541
37 Rue Collange, 92300 LeVallois, Perret, France • Telephone: (33) 1-427-02642 • Fax: (33) 1-427-02798
44 Wilby Avenue, Little Lever, Bolton, Lancashire, BL31QE, U.K. • Telephone: (44) 120-457-6068 • Fax: (44) 120-479-6463
E-mail: substrates@arlonmed.com • Website: www.arlonmed.com

Arlon is an ISO 9002 Registered Company

Microwave Materials

PTFE/Woven Fiberglass/Ceramic Filled Laminates: Microwave Printed Circuit Board Substrates

AR600™ is a woven fiberglass, reinforced ceramic filled PTFE based composite material for use as a printed circuit board substrate.

The higher dielectric constant of AR600 permits moderate circuit miniaturization, especially for lower frequency microwave and power amplifier applications which use low impedance lines.

AR600 is a "soft substrate" and is not sensitive to vibrational stress. This allows miniaturized circuitry without requiring the complicated processing or fragile handling associated with a brittle pure ceramic material.

AR600 is compatible with the processing used for standard PTFE based printed circuit board substrates. In addition, the low Z-axis thermal expansion provided by the ceramic loading will improve plated through hole reliability compared to typical PTFE based laminates.

Availability:

AR600 laminates are supplied with 1/2, 1 or 2 ounce electrodeposited copper on both sides. Other copper weights and rolled copper foil are available. AR600 is available bonded to a heavy metal ground plane. Aluminum, brass or copper plates also provide an integral heat sink and mechanical support to the substrate.

When ordering AR600 products, please specify thickness, cladding, panel size, and any other special considerations. Available master sheet sizes include 36" x 48" and 36" x 72" (48" x 54" master sheet size pending).

Typical Properties: AR600™ PTFE/Woven Fiberglass/Ceramic Filled Laminates

Properties	Test Method	Condition	Typical Values
Dielectric Constant @10GHz	IPC TM-650 2.5.5.6	C23/50	6.0
Dissipation Factor @10GHz	IPC TM-650 2.5.5.6	C23/50	0.0035
Thermal Coefficient of E_r (ppm/°C)	IPC TM-650 2.5.5.5 Adapted	-10°C to +140°C	-325
Volume Resistivity (MΩ-cm)	IPC TM-650 2.5.17.1	C96/35/90	1.5×10^{12}
Surface Resistivity (MΩ)	IPC TM-650 2.5.17.1	C96/35/90	3.8×10^{9}
Arc Resistance (seconds)	ASTM D-495	D48/50	> 180
Tensile Modulus (kpsi)	ASTM D-638	A, 23°C	700, 500
Tensile Strength (kpsi)	ASTM D-882	A, 23°C	8.3, 7.0
Compressive Modulus (kpsi)	ASTM D-695	A, 23°C	225
Flexural Modulus (kpsi)	ASTM D-790	A, 23°C	375
Dielectric Breakdown (kv)	ASTM D-149	D48/50	> 45
Specific Gravity (g/cm³)	ASTM D-792 Method A	A, 23°C	2.45
Water Absorption (%)	MIL-S-13949H 3.7.7 IPC TM-650 2.6.2.2	E1/105 + D24/23	0.08
Coefficient of Thermal Expansion (ppm/°C)	IPC TM-650 2.4.24 Mettler 3000 Thermomechanical Analyzer	0°C to 100°C	
X Axis			12
Y Axis			14
Z Axis			62
Thermal Conductivity (W/mK)	ASTM E-1225	100°C	0.431
Outgassing	NASA SP-R-0022A		
Total Mass Loss (%)	Maximum 1.00%	125°C, ≤ 10^{-6} torr	0.02
Collected Volatile Condensable Material (%)	Maximum 0.10%		0.00
Water Vapor Regain (%)			0.00
Visible Condensate (±)			NO
Flammability	UL 94 Vertical Burn IPC TM-650 2.3.10	C48/23/50, E24/125	Meets requirements of UL94V-0

Data based on 0.062" dielectric thickness, exclusive of metal cladding except where indicated by test method. Results listed above are typical properties; they are not to be used as specification limits. The above information creates no expressed or implied warranties. The properties of AR600 laminates may vary depending on the application.

The information and data contained herein are believed reliable, but all recommendations or suggestions are made without guarantee. You should thoroughly and independently test materials for any planned applications and determine satisfactory performance before commercialization. Furthermore, no suggestion for use, or material supplied shall be construed as a recommendation or inducement to violate any law or infringe any patent.

MATERIALS FOR ELECTRONICS

1100 Governor Lea Road, Bear, DE 19701 • Telephone: (302) 834-2100, (800) 635-9333 • Fax: (302) 834-2574
9433 Hyssop Drive, Rancho Cucamonga, CA 91730 • Telephone: (909) 987-9533 • Fax: (909) 987-8541
37 Rue Collange, 92300 LeVallois, Perret, France • Telephone: (33) 1-427-02642 • Fax: (33) 1-427-02798
44 Wilby Avenue, Little Lever, Bolton, Lancashire, BL31QE, U.K. • Telephone: (44) 120-457-6068 • Fax: (44) 120-479-6463
E-mail: substrates@arlonmed.com • Website: www.arlonmed.com

Arlon is an ISO 9002 Registered Company

Microwave Materials

PTFE/Woven Fiberglass/Ceramic Filled Laminates: Microwave Printed Circuit Board Substrates

AR1000® is a woven fiberglass, reinforced ceramic filled PTFE based composite material for use as a printed circuit board substrate when a high Dk (10.0) is necessary for circuit optimization.

The higher dielectric constant of AR1000 permits moderate circuit miniaturization, especially for lower frequency microwave and power amplifier applications which use low impedance lines.

AR1000 is a "soft substrate" and is not sensitive to vibrational stress. This allows miniaturized circuitry without requiring the complicated processing or fragile handling associated with a brittle pure ceramic material.

AR1000 is compatible with the processing used for standard PTFE based printed circuit board substrates. In addition, the low Z-axis thermal expansion provided by the ceramic loading will improve plated through hole reliability compared to typical PTFE based laminates.

Availability:

AR1000 laminates are supplied with 1/2, 1 or 2 ounce electrodeposited copper on both sides. Other copper weights and rolled copper foil are available. AR1000 is available bonded to a heavy metal ground plane. Aluminum, brass or copper plates also provide an integral heat sink and mechanical support to the substrate.

When ordering AR1000 products, please specify thickness, cladding, panel size and any other special considerations. Available master sheet sizes include 36" x 48" and 36" x 72" (48" x 54" master sheet size pending).

Typical Properties: AR1000® PTFE/Woven Fiberglass/Ceramic Filled Laminates

Properties	Test Method	Condition	Typical Values
Dielectric Constant @10GHz	IPC TM-650 2.5.5.6	C23/50	10.0 (varies by thickness)
Dissipation Factor @10GHz	IPC TM-650 2.5.5.6	C23/50	0.003
Thermal Coefficient of E_r (ppm/°C)	IPC TM-650 2.5.5.5 Adapted	-10°C to +140°C	-233
Volume Resistivity (MΩ-cm)	IPC TM-650 2.5.17.1	C96/35/90	1.4×10^9
Surface Resistivity (MΩ)	IPC TM-650 2.5.17.1	C96/35/90	1.8×10^9
Arc Resistance (seconds)	ASTM D-495	D48/50	> 180
Tensile Modulus (kpsi)	ASTM D-638	A, 23°C	830, 680
Tensile Strength (kpsi)	ASTM D-882	A, 23°C	5.1, 4.3
Compressive Modulus (kpsi)	ASTM D-695	A, 23°C	450
Flexural Modulus (kpsi)	ASTM D-790	A, 23°C	615
Dielectric Breakdown (kv)	ASTM D-149	D48/50	> 45
Specific Gravity (g/cm³)	ASTM D-792 Method A	A, 23°C	2.84
Water Absorption (%)	MIL-S-13949H 3.7.7 IPC TM-650 2.6.2.2	E1/105 + D24/23	0.08
Coefficient of Thermal Expansion (ppm/°C)	IPC TM-650 2.4.24 Mettler 3000 Thermomechanical Analyzer	0°C to 100°C	
X Axis			14
Y Axis			16
Z Axis			36
Thermal Conductivity (W/mK)	ASTM E-1225	100°C	0.645
Outgassing	NASA SP-R-0022A		
Total Mass Loss (%)	Maximum 1.00%	125°C, ≤ 10^{-6} torr	0.02
Collected Volatile Condensable Material (%)	Maximum 0.10%		0.00
Water Vapor Recovered (%)			0.00
Visible Condensate (±)			NO
Flammability	UL 94 Vertical Burn IPC TM-650 2.3.10	C48/23/50, E24/125	Meets requirements of UL94V-0

Data based on 0.062" dielectric thickness, exclusive of metal cladding except where indicated by test method. Results listed above are typical properties; they are not to be used as specification limits. The above information creates no expressed or implied warranties. The properties of AR1000 laminates may vary depending on the application.

The information and data contained herein are believed reliable, but all recommendations or suggestions are made without guarantee. You should thoroughly and independently test materials for any planned applications and determine satisfactory performance before commercialization. Furthermore, no suggestion for use, or material supplied shall be construed as a recommendation or inducement to violate any law or infringe any patent.

1100 Governor Lea Road, Bear, DE 19701 • Telephone: (302) 834-2100, (800) 635-9333 • Fax: (302) 834-2574
9433 Hyssop Drive, Rancho Cucamonga, CA 91730 • Telephone: (909) 987-9533 • Fax: (909) 987-8541
37 Rue Collange, 92300 LeVallois, Perret, France • Telephone: (33) 1-427-02642 • Fax: (33) 1-427-02798
44 Wilby Avenue, Little Lever, Bolton, Lancashire, BL31QE, U.K. • Telephone: (44) 120-457-6068 • Fax: (44) 120-479-6463
E-mail: substrates@arlonmed.com • Website: www.arlonmed.com

Arlon is an ISO 9002 Registered Company

Microwave Materials

PTFE/Nonwoven Fiberglass Laminates: Microwave Printed Circuit Board Substrates

IsoClad® laminates are nonwoven fiberglass/PTFE composites for use as printed circuit board substrates. The nonwoven reinforcement allows these laminates to be used more easily in applications where the final circuit will be bent to shape. Conformal or "wrap-around" antennas are a good example.

IsoClad products use longer random fibers and a proprietary process to provide greater dimensional stability and better dielectric constant uniformity than competitive nonwoven fiberglass/PTFE laminates of similar dielectric constants.

IsoClad 917 (ϵ_r=2.17, 2.20) uses a low ratio of fiberglass/PTFE to achieve the lowest dielectric constant and dissipation factor available in a combination of PTFE and fiberglass.

IsoClad 933 (ϵ_r=2.33) uses a higher fiberglass/PTFE ratio for a more highly reinforced combination which offers better dimensional stability and increased mechanical strength.

Availability:

IsoClad laminates are supplied with 1/2, 1 or 2 ounce electrodeposited copper on both sides. Other copper weights and rolled copper foil are available. IsoClad is available bonded to a heavy metal ground plane. Aluminum, brass or copper plates also provide an integral heat sink and mechanical support to the substrate.

When ordering IsoClad products, please specify dielectric constant, thickness, cladding, panel size and any other special considerations. Available master sheet sizes include 36" x 48" and 36" x 72".

Typical Properties: IsoClad® PTFE/Nonwoven Fiberglass Laminates

Properties	Test Method	Condition	Typical Values IsoClad 917	Typical Values IsoClad 933
Dielectric Constant @10GHz	IPC TM-650 2.5.5.5	C23/50	2.17, 2.20	2.33
Dissipation Factor @10GHz	IPC TM-650 2.5.5.5	C23/50	0.0013	0.0016
Thermal Coefficient of E_r (ppm/°C)	IPC TM-650 2.5.5.5 Adapted	-10°C to +140°C	-157	-132
Volume Resistivity (MΩ-cm)	IPC TM-650 2.5.17.1	C96/35/90	2.4×10^{14}	3.9×10^{13}
Surface Resistivity (MΩ)	IPC TM-650 2.5.17.1	C96/35/90	3.2×10^{12}	1.1×10^{11}
Arc Resistance (seconds)	ASTM D-495	D48/50	> 180	> 180
Tensile Modulus (kpsi)	ASTM D-638	A, 23°C	133, 120	173, 147
Tensile Strength (kpsi)	ASTM D-882	A, 23°C	4.3, 3.8	6.8, 5.3
Compressive Modulus (kpsi)	ASTM D-695	A, 23°C	182	197
Flexural Modulus (kpsi)	ASTM D-790	A, 23°C	213	239
Dielectric Breakdown (kv)	ASTM D-149	D48/50	> 45	> 45
Specific Gravity (g/cm³)	ASTM D-792 Method A	A, 23°C	2.23	2.27
Water Absorption (%)	MIL-S-13949H 3.7.7 IPC TM-650 2.6.2.2	E1/105 + D24/23	0.04	0.05
Coefficient of Thermal Expansion (ppm/°C)	IPC TM-650 2.4.24 Mettler 3000 Thermomechanical Analyzer	0°C to 100°C		
X Axis			46	31
Y Axis			47	35
Z Axis			236	203
Thermal Conductivity (W/mK)	ASTM E-1225	100°C	0.263	0.263
Outgassing	NASA SP-R-0022A			
Total Mass Loss (%)	Maximum 1.00%	125°C, ≤ 10^{-6} torr	0.02	0.03
Collected Volatile Condensable Material (%)	Maximum 0.10%		0.00	0.00
Water Vapor Regain (%)			0.02	0.02
Visible Condensate (±)			NO	NO
Flammability	UL 94 Vertical Burn IPC TM-650 2.3.10	C48/23/50, E24/125	Meets requirements of UL94V-0	Meets requirements of UL94V-0

Data based on 0.062" dielectric thickness, exclusive of metal cladding except where indicated by test method. Results listed above are typical properties; they are not to be used as specification limits. The above information creates no expressed or implied warranties. The properties of IsoClad laminates may vary depending on the application.

The information and data contained herein are believed reliable, but all recommendations or suggestions are made without guarantee. You should thoroughly and independently test materials for any planned applications and determine satisfactory performance before commercialization. Furthermore, no suggestion for use, or material supplied shall be construed as a recommendation or inducement to violate any law or infringe any patent.

MATERIALS FOR ELECTRONICS

1100 Governor Lea Road, Bear, DE 19701 • Telephone: (302) 834-2100, (800) 635-9333 • Fax: (302) 834-2574
9433 Hyssop Drive, Rancho Cucamonga, CA 91730 • Telephone: (909) 987-9533 • Fax: (909) 987-8541
37 Rue Collange, 92300 LeVallois, Perret, France • Telephone: (33) 1-427-02642 • Fax: (33) 1-427-02798
44 Wilby Avenue, Little Lever, Bolton, Lancashire, BL31QE, U.K. • Telephone: (44) 120-457-6068 • Fax: (44) 120-479-6463
E-mail: substrates@arlonmed.com • Website: www.arlonmed.com

Arlon is an ISO 9002 Registered Company

Microwave Materials

Dimensionally and Electrically Stable Microwave Printed Circuit Board Substrates

CLTE® is a ceramic powder-filled and woven microfiberglass reinforced PTFE composite engineered to produce a stable, low water absorption laminate with a nominal dielectric constant of 2.94.

Arlon's proprietary formulation for CLTE materials creates a reduced Z-direction thermal expansion (nearer to the expansion rate for copper metal), improving plated through hole reliability. It is stable during subsequent thermal cycling in process, assembly and use.

The formulation was chosen to minimize the change in ϵ_r caused by the PTFE 19° C second order phase transition of the molecular structure. This temperature stable ϵ_r simplifies circuit design and optimizes circuit performance in applications such as phased array antennas.

CLTE also provides higher thermal conductivity which increases the rate of heat dissipation and thus permits use of higher power in an otherwise equivalent design.

CLTE retains the low loss tangent associated with PTFE. While once required only for microwave frequencies, low loss is also of great value in reducing cross talk in high-speed digital applications and minimizes the power consumption of a circuit design.

Two grades of thermoplastic prepregs are available to match the stable electrical and mechanical performance characteristics of CLTE laminates. CLTE-P and CLTE-PC offer two different lamination temperature options for multilayer applications. See page 4 for laminating recommendations.

Availability:

CLTE laminates are supplied with 1/2, 1 or 2 ounce electrodeposited copper on both sides. Other copper weights and rolled copper foil are available. CLTE is available bonded to a heavy metal ground plane. Aluminum, brass or copper plates also provide an integral heat sink and mechanical support to the substrate.

When ordering CLTE products please specify thickness, cladding, panel size and any other special considerations. Available master sheet sizes include 36" x 48", 36" x 72" and 48" x 54".

Typical Properties: CLTE® Dimensionally and Electrically Stable Microwave Substrate Materials

Properties	Test Method	Condition	Typical Values
Dielectric Constant @10GHz	IPC TM-650 2.5.5.5	C23/50	2.94
Dissipation Factor @10GHz	IPC TM-650 2.5.5.5	C23/50	0.0025
Thermal Coefficient of E_r (ppm/°C)	IPC TM-650 2.5.5.5 Adapted	-10°C to +140°C	See graph
Volume Resistivity (MΩ-cm)	IPC TM-650 2.5.17.1	C96/35/90	1.4×10^8
Surface Resistivity (MΩ)	IPC TM-650 2.5.17.1	C96/35/90	1.3×10^6
Arc Resistance (seconds)	ASTM D-495	D48/50	> 180
Tensile Modulus (kpsi)	ASTM D-638	A, 23°C	471, 462
Tensile Strength (kpsi)	ASTM D-882	A, 23°C	8.2, 7.0
Compressive Modulus (kpsi)	ASTM D-695	A, 23°C	225
Flexural Modulus (kpsi)	ASTM D-790	A, 23°C	375
Dielectric Breakdown (kv)	ASTM D-149	D48/50	> 45
Specific Gravity (g/cm³)	ASTM D-792 Method A	A, 23°C	2.38
Water Absorption (%)	MIL-S-13949H 3.7.7 IPC TM-650 2.6.2.2	E1/105 + D24/23	0.04
Coefficient of Thermal Expansion (ppm/°C)	IPC TM-650 2.4.24 Mettler 3000 Thermomechanical Analyzer	0°C to 100°C	
X Axis			10
Y Axis			12
Z Axis			40
Thermal Conductivity (W/mK)	ASTM E-1225	100°C	0.50
Outgassing	NASA SP-R-0022A		
Total Mass Loss (%)	Maximum 1.00%	125°C, ≤ 10^{-6} torr	0.02
Collected Volatile Condensable Material (%)	Maximum 0.10%		0.00
Water Vapor Recovered (%)			0.00
Visible Condensate (±)			NO
Flammability (UL File E 80166)	UL 94 Vertical Burn IPC TM-650 2.3.10	C48/23/50, E24/125	UL94V-0

Data based on 0.062" dielectric thickness, exclusive of metal cladding except where indicated by test method. Results listed above are typical properties; they are not to be used as specification limits. The above information creates no expressed or implied warranties. The properties of CLTE laminates may vary depending on the application.

MATERIALS FOR ELECTRONICS

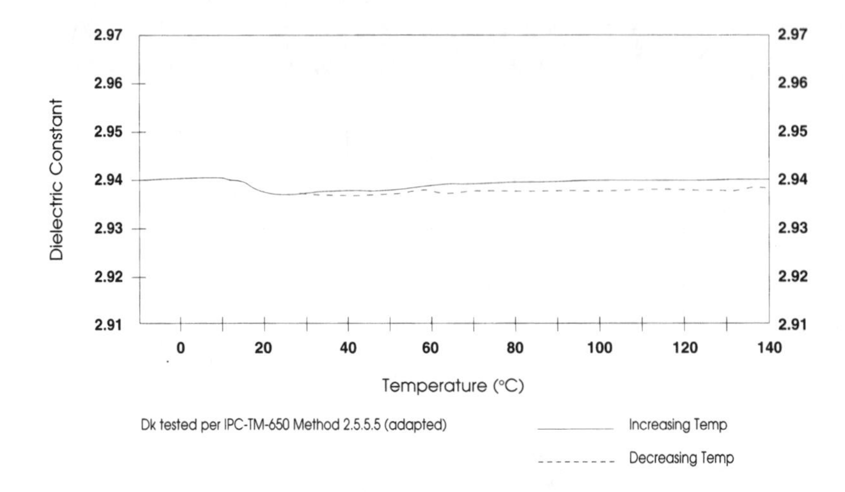

THIS DK/TEMPERATURE CURVE shows the unique thermal stability properties of CLTE materials when thermocycled over temperature. Even over a wide temperature variation, the material retains its ultra-stable dielectric constant characteristics. Because of its low thermal expansion properties, CLTE material is ideal for the fabrication of complex multilayer circuits (see page four for laminating recommendations).

Multilayer Lamination Recommendations

Following the use of conventional imaging and etching processes, successful fabrication of multilayer circuit assemblies using the CLTE Series pre-pregs (designated CLTE-P and CLTE-PC) with the CLTE series laminates can be achieved through use of the following recommendations.

1. Pre-preg Material

The Pre-preg material consists of woven fiberglass fabric coated with a proprietary resin formulation. As received the thickness of pre-preg is ≈.0032". After lamination, the thickness is compressed to ≈.0024".

2. Surface Preparation

a. ***Substrate surface*** - No additional surface treatment, either mechanical or chemical, should be necessary to achieve good adhesion. However, this recommendation is based upon laboratory conditions where multilayer lamination was performed immediately after etching of the copper surface. For panels which have a long wait time between etching and lamination, a sodium etch (or plasma etch process appropriate for PTFE) of the CLTE laminate surface will provide optimal results.

b. ***Copper surfaces*** - Microetch and dry the inner layer copper surfaces immediately prior to lay-up and lamination. Standard FR-4 black oxide processes may not provide optimal results due to the high lamination temperatures required to bond CLTE-P and CLTE-PC. Brown or red oxide treatments may improve the bond to large copper plane areas.

3. Lamination

a. ***Equipment*** - A press which has heat and cool cycles in the same opening is recommended. This ensures that constant pressure can be maintained throughout both the heat-up and cool-down cycle.

b. ***Temperature*** - CLTE-PC requires a lamination temperature of 450°F/232°C to allow sufficient flow of the resin; CLTE-P requires a lamination temperature of 550°F/288°C to allow sufficient flow of the resin. The lamination temperature should be measured at the bond line using a thermocouple located in the edge of the product panel.

Because of the high temperatures required for lamination, noncombustible peripheral materials, such as separator sheets and press padding material, should be used. Epoxy separator sheets are not recommended as they may char or burn. Paper and certain rubber press padding materials are also not recommended for similar reasons.

c. ***Pressure (400 psi actual)*** - A pressure of 400 psi is recommended to remove any entrapped air and force the flow of the pre-preg into the exposed "tooth" present on the surface of the laminate. This pressure must be maintained throughout the full extent of the heating and cooling cycles.

d. ***Heat up and cool down rate*** - Since CLTE-P and CLTE-PC are thermoplastic materials, precise control of heat up and cool down rates is not critical.

e. ***Time at laminating temperature (45 minutes)*** - Good adhesion will be achieved by maintaining the recommended laminating temperature for a period of 45 minutes.

MATERIALS FOR ELECTRONICS

1100 Governor Lea Road, Bear, DE 19701 • Telephone: (302) 834-2100, (800) 635-9333 • Fax: (302) 834-2574
9433 Hyssop Drive, Rancho Cucamonga, CA 91730 • Telephone: (909) 987-9533 • Fax: (909) 987-8541
37 Rue Collange, 92300 LeVallois, Perret, France • Telephone: (33) 1-427-02642 • Fax: (33) 1-427-02798
44 Wilby Avenue, Little Lever, Bolton, Lancashire, BL31QE, U.K. • Telephone: (44) 120-457-6068 • Fax: (44) 120-479-6463
E-mail: substrates@arlonmed.com • Website: www.arlonmed.com

Arlon is an ISO 9002 Registered Company

Microwave Materials

PTFE/Woven Fiberglass Laminates: Microwave Printed Circuit Board Substrates

DiClad® laminates are woven fiberglass/PTFE composite materials for use as printed circuit board substrates. Using precise control of the fiberglass/PTFE ratio, DiClad laminates offer a range of choices from the lowest dielectric constant and dissipation factor to a more highly reinforced laminate with better dimensional stability.

The woven fiberglass reinforcement in DiClad products provides greater dimensional stability than nonwoven fiberglass reinforced PTFE based laminates of similar dielectric constants. The consistency and control of the PTFE coated fiberglass cloth allows Arlon to offer a greater variety of dielectric constants and produces a laminate with better dielectric constant uniformity than comparable nonwoven fiberglass reinforced laminates. The coated fiberglass plies in DiClad materials are aligned in the same direction. Crossplied versions of many of these materials are available as Arlon CuClad materials.

DiClad laminates are frequently used in filter, coupler and low noise amplifier applications, where dielectric constant uniformity is critical. They are also used in power dividers and combiners where low loss is important.

DiClad 522 and DiClad 527 (ϵ_r=2.40–2.65) use a higher fiberglass/PTFE ratio to provide mechanical properties approaching conventional substrates. Other advantages include better dimensional stability and lower thermal expansion in all directions. The electrical properties of DiClad 522 and DiClad 527 are tested at 1MHz and 10GHz respectively.

DiClad 870 (ϵ_r=2.33) uses a medium fiberglass/PTFE ratio for lower dielectric constant and improved dissipation factor without sacrificing mechanical properties.

DiClad 880 (ϵ_r=2.17, 2.20) uses a low fiberglass/PTFE ratio to provide the lowest dielectric constant and dissipation factor available in fiberglass reinforced PTFE based laminates. Together, these properties offer faster signal propagation and higher signal to noise ratios.

Availability:

DiClad laminates are supplied with 1/2, 1 or 2 ounce electrodeposited copper on both sides. Other copper weights and rolled copper foil are available. DiClad is available bonded to a heavy metal ground plane. Aluminum, brass or copper plates also provide an integral heat sink and mechanical support to the substrate.

When ordering DiClad products please specify dielectric constant, thickness, cladding, panel size and any other special considerations. Available master sheet sizes include 36" x 48", 36" x 72" and 48" x 54".

Typical Properties: DiClad® PTFE/Woven Fiberglass Laminates

Properties	Test Method	Condition	Typical Values DiClad 880	Typical Values DiClad 870	Typical Values DiClad 522/527
Dielectric Constant @10GHz	IPC TM-650 2.5.5.5	C23/50	2.17, 2.20	2.33	2.40 to 2.65
Dielectric Constant @1MHz	IPC TM-650 2.5.5.3	C23/50	2.17, 2.20	2.33	2.40 to 2.65
Dissipation Factor @10GHz	IPC TM-650 2.5.5.5	C23/50	0.0009	0.0013	0.0022
Dissipation Factor @1MHz	IPC TM-650 2.5.5.3	C23/50	0.0008	0.0009	0.0010
Thermal Coefficient of E_r (ppm/°C)	IPC TM-650 2.5.5.5 Adapted	-10°C to +140°C	-160	-161	-153
Volume Resistivity (MΩ-cm)	IPC TM-650 2.5.17.1	C96/35/90	1.4×10^9	1.5×10^9	1.2×10^9
Surface Resistivity (MΩ)	IPC TM-650 2.5.17.1	C96/35/90	2.9×10^6	3.4×10^7	4.5×10^7
Arc Resistance (seconds)	ASTM D-495	D48/50	> 180	> 180	> 180
Tensile Modulus (kpsi)	ASTM D-638	A, 23°C	267, 202	485, 346	706, 517
Tensile Strength (kpsi)	ASTM D-882	A, 23°C	8.1, 7.5	14.9, 11.2	19.0, 15.0
Compressive Modulus (kpsi)	ASTM D-695	A, 23°C	237	327	359
Flexural Modulus (kpsi)	ASTM D-790	A, 23°C	357	437	537
Dielectric Breakdown (kv)	ASTM D-149	D48/50	> 45	> 45	> 45
Specific Gravity (g/cm³)	ASTM D-792 Method A	A, 23°C	2.23	2.26	2.31
Water Absorption (%)	MIL-S-13949H 3.7.7 IPC TM-650 2.6.2.2	E1/105 + D24/23	0.02	0.02	0.03
Coefficient of Thermal Expansion (ppm/°C)	IPC TM-650 2.4.24 Mettler 3000 Thermomechanical Analyzer	0°C to 100°C			
X Axis			25	17	14
Y Axis			34	29	21
Z Axis			252	217	173
Thermal Conductivity (W/mK)	ASTM E-1225	100°C	0.261	0.257	0.254
Outgassing	NASA SP-R-0022A				
Total Mass Loss (%)	Maximum 1.00%	125°C, ≤ 10^{-6} torr	0.01	0.02	0.02
Collected Volatile Condensable Material (%)	Maximum 0.10%		0.00	0.00	0.00
Water Vapor Regain (%)			0.00	0.00	0.00
Visible Condensate (±)			NO	NO	NO
Flammability UL File E 80166	UL 94 Vertical Burn IPC TM-650 2.3.10	C48/23/50, E24/125	UL94V-0	UL94V-0	UL94V-0

Data based on 0.062" dielectric thickness, exclusive of metal cladding except where indicated by test method. Results listed above are typical properties; they are not to be used as specification limits. The above information creates no expressed or implied warranties. The properties of DiClad laminates may vary depending on the application.

The information and data contained herein are believed reliable, but all recommendations or suggestions are made without guarantee. You should thoroughly and independently test materials for any planned applications and determine satisfactory performance before commercialization. Furthermore, no suggestion for use, or material supplied shall be construed as a recommendation or inducement to violate any law or infringe any patent.

MATERIALS FOR ELECTRONICS

1100 Governor Lea Road, Bear, DE 19701 • Telephone: (302) 834-2100, (800) 635-9333 • Fax: (302) 834-2574
9433 Hyssop Drive, Rancho Cucamonga, CA 91730 • Telephone: (909) 987-9533 • Fax: (909) 987-8541
37 Rue Collange, 92300 LeVallois, Perret, France • Telephone: (33) 1-427-02642 • Fax: (33) 1-427-02798
44 Wilby Avenue, Little Lever, Bolton, Lancashire, BL31QE, U.K. • Telephone: (44) 120-457-6068 • Fax: (44) 120-479-6463
E-mail: substrates@arlonmed.com • Website: www.arlonmed.com

Arlon is an ISO 9002 Registered Company

Microwave Materials

PTFE/Woven Fiberglass Laminates: Microwave Printed Circuit Board Substrates

CuClad® laminates are woven fiberglass/PTFE composite materials for use as printed circuit board substrates. Using precision control of the fiberglass/PTFE ratio, CuClad laminates offer a range of choices from the lowest dielectric constant and loss tangent to a more highly reinforced laminate with better dimensional stability.

The woven fiberglass reinforcement in CuClad products provides greater dimensional stability than non-woven fiberglass reinforced PTFE based laminates of similar dielectric constants. The consistency and control of the PTFE coated fiberglass cloth allows Arlon to offer a greater variety of dielectric constants and produces a laminate with better dielectric constant uniformity than comparable non-woven fiberglass reinforced laminates. These properties make CuClad an attractive choice for filters, couplers and low noise amplifiers.

CuClad laminates are crossplied (alternating layers of coated fiberglass plies are oriented 90° to each other). This provides true electrical and mechanical isotropy in the XY plane, a feature unique to CuClad. No other woven or nonwoven fiberglass reinforced PTFE based laminates make this claim. Designers have found this degree of isotropy critical in some phased array antenna applications.

CuClad 217 (ϵ_r=2.17, 2.20) uses a low fiberglass/PTFE ratio to provide the lowest dielectric constant and dissipation factor available in fiberglass reinforced PTFE based laminates. Together, these properties offer faster signal propagation and higher signal/noise ratios.

CuClad 233 (ϵ_r=2.33) uses a medium fiberglass/PTFE ratio to balance lower dielectric constant and improved dissipation factor without sacrificing mechanical properties.

CuClad 250 (ϵ_r=2.40–2.60) uses a higher fiberglass/PTFE ratio to provide mechanical properties approaching those of conventional substrates. Better dimensional stability and lower thermal expansion in all directions are other significant benefits. The electrical properties of CuClad 250GT and CuClad 250GX are tested at 1MHz and 10GHz respectively.

For critical performance applications, CuClad products may be specified with an "LX" testing grade; this designates that each sheet will be tested individually, and a test report will be issued with the order

Availability:

CuClad laminates are supplied with 1/2, 1 or 2 ounce electrodeposited copper on both sides. Other copper weights and rolled copper foil are available. CuClad is available bonded to a heavy metal ground plane. Aluminum, brass or copper plates also provide an integral heat sink and mechanical support to the substrate.

When ordering CuClad products please specify dielectric constant, thickness, cladding, panel size and any other special considerations. Available master sheet sizes include 36" x 36" in a crossplied configuration and 36" x 48" in a parallel plied configuration.

MATERIALS FOR ELECTRONICS

Typical Properties: CuClad® PTFE/Woven Fiberglass Laminates

Properties	Test Method	Condition	Typical Values CuClad 217	Typical Values CuClad 233	Typical Values CuClad 250
Dielectric Constant @10GHz	IPC TM-650 2.5.5.5	C23/50	2.17, 2.20	2.33	2.40 to 2.60
Dielectric Constant @1MHz	IPC TM-650 2.5.5.3	C23/50	2.17, 2.20	2.33	2.40 to 2.60
Dissipation Factor @10GHz	IPC TM-650 2.5.5.5	C23/50	0.0009	0.0013	0.0022
Dissipation Factor @1MHz	IPC TM-650 2.5.5.3	C23/50	0.0008	0.0009	0.0010
Thermal Coefficient of E_r (ppm/°C)	IPC TM-650 2.5.5.5 Adapted	-10°C to +140°C	-151	-171	-170
Volume Resistivity (MΩ-cm)	IPC TM-650 2.5.17.1	C96/35/90	2.3×10^8	8.0×10^8	1.8×10^9
Surface Resistivity (MΩ)	IPC TM-650 2.5.17.1	C96/35/90	3.4×10^6	2.4×10^6	1.5×10^8
Arc Resistance (seconds)	ASTM D-495	D48/50	> 180	> 180	> 180
Tensile Modulus (kpsi)	ASTM D-638	A, 23°C	275, 219	510, 414	725, 572
Tensile Strength (kpsi)	ASTM D-882	A, 23°C	8.8, 6.6	10.3, 9.8	26.0, 20.5
Compressive Modulus (kpsi)	ASTM D-695	A, 23°C	237	276	342
Flexural Modulus (kpsi)	ASTM D-790	A, 23°C	357	371	456
Dielectric Breakdown (kv)	ASTM D-149	D48/50	> 45	> 45	> 45
Specific Gravity (g/cm³)	ASTM D-792 Method A	A, 23°C	2.23	2.26	2.31
Water Absorption (%)	MIL-S-13949H 3.7.7 IPC TM-650 2.6.2.2	E1/105 + D24/23	0.02	0.02	0.03
Coefficient of Thermal Expansion (ppm/°C)	IPC TM-650 2.4.24 Mettler 3000 Thermomechanical Analyzer	0°C to 100°C			
X Axis			29	23	18
Y Axis			28	24	19
Z Axis			246	194	177
Thermal Conductivity (W/mK)	ASTM E-1225	100°C	0.261	0.258	0.254
Outgassing	NASA SP-R-0022A				
Total Mass Loss (%)	Maximum 1.00%	125°C, ≤ 10^{-6} torr	0.01	0.01	0.01
Collected Volatile Condensable Material (%)	Maximum 0.10%		0.00	0.00	0.00
Water Vapor Regain (%)			0.00	0.00	0.00
Visible Condensate (±)			NO	NO	NO
Flammability UL File E 80166	UL 94 Vertical Burn IPC TM-650 2.3.10	C48/23/50, E24/125	UL94V-0	UL94V-0	UL94V-0

Data based on 0.062" dielectric thickness, exclusive of metal cladding except where indicated by test method. Results listed above are typical properties; they are not to be used as specification limits. The above information creates no expressed or implied warranties. The properties of CuClad Series laminates may vary depending on the application.

The information and data contained herein are believed reliable, but all recommendations or suggestions are made without guarantee. You should thoroughly and independently test materials for any planned applications and determine satisfactory performance before commercialization. Furthermore, no suggestion for use, or material supplied shall be construed as a recommendation or inducement to violate any law or infringe any patent.

MATERIALS FOR ELECTRONICS

1100 Governor Lea Road, Bear, DE 19701 • Telephone: (302) 834-2100, (800) 635-9333 • Fax: (302) 834-2574
9433 Hyssop Drive, Rancho Cucamonga, CA 91730 • Telephone: (909) 987-9533 • Fax: (909) 987-8541
37 Rue Collange, 92300 LeVallois, Perret, France • Telephone: (33) 1-427-02642 • Fax: (33) 1-427-02798
44 Wilby Avenue, Little Lever, Bolton, Lancashire, BL31QE, U.K. • Telephone: (44) 120-457-6068 • Fax: (44) 120-479-6463
E-mail: substrates@arlonmed.com • Website: www.arlonmed.com

Arlon is an ISO 9002 Registered Company

MC3 Polyester/Glass Laminate

Typical Properties of MC3 (Single Sided)

Description
Grade MC3 is a composite laminate targeted for the rigid, single-sided market. Comprised of woven glass surface sheets on either side of a glass paper core, MC3 can be punched cleanly with a minimum amount of tool wear. The unique resin system is compatible with conventional PWB chemistries. In addition, MC3 exhibits excellent electricals with a low and stable dielectric constant/dissipation factor and a high Comparative Tracking Index. MC3 is an excellent material for high frequency or high voltage applications.

Features and Benefits

- Fabrication in conventional PWB (FR4) shops
- Standard FR4 drilling / cutting feeds and speeds.
- Conventional glass construction
- No special storage requirements
- Standard assembly techniques.
- UL 94 V-0 flammability rating
- **Absolutely the best cost performance available**

UL Information

File .... #E87492
Grade MC3
Max Operating Temperature
Electrical .... 140°C
Mechanical .... 130°C
Flame Class 94V-0
Solder Limits
0.031" .... 525°F, 20 sec
0.062" .... 550°F, 30 sec

Electrical Properties			
Dielectric Constant	@ 1 Mhz	3.7	
Dissipation Factor	@ 1 MHZ	0.018	
Surface Resistivity	C-96/35/90	Megohm	1 X 10^9
Volume Resistivity	C-96/35/90	Megohm-cm	5 X 10^8
Arc Resistance	A	seconds	120
Dielectric Breakdown	D-48/50	kV	70
Comparative Tracking Index	A	Volts	600+
Physical Properties			
Copper Peel Strength	A	lb / inch width	8.0
	After 20 sec @500°F	lb / inch width	8.0
Flex Strength Length	A	psi	43,000
Flex Strength Cross	A	psi	26,500
Water Absorption	D-24/23	%	0.14
Thermal Properties			
Glass Transition (Tg)	A	°C	95
Thermal Stress	@ 500 °F	seconds	40+
Z-Axis CTE, RT → Tg	A	ppm/°C	200
Z-Axis CTE, Tg → 260°C	A	ppm/°C	400
Dimensional Stability, Length	E-4/105, E-2/150	inch/inch	-0.00034
Dimensional Stability, Cross	E-4/105, E-2/150	inch/inch	-0.00055

- Typical Properties of 0.059" thick laminate, clad with 1 ounce copper. Properties of other thicknesses may vary.
- All test methods are IPC-TM-650
- Tg measured by DMA

Rev G **TSR-133**

175 Commerce Road • Collierville, TN 38017 • 901-853-5070 • Fax 901-853-8664 • gilam@gilam.com

MC3D Medium Frequency Laminate

Typical Properties of MC3D Double Sided (0.031 inch)

Description
Grade MC3D is a composite laminate targeted for the rigid, double-sided market. Comprised of woven glass surface sheets on either side of a glass paper core, MC3D can be drilled cleanly with a minimum amount of drill wear. The unique resin system is compatible with conventional etchback and plated through hole chemistries. In addition, MC3D exhibits excellent electricals with a low and stable dielectric constant and dissipation factor. MC3D is an excellent material for frequency sensitive applications.

Features and Benefits
- Fabrication in conventional PWB (FR4) shops
- Standard FR4 drilling / cutting feeds and speeds.
- No special through hole treatments
- Conventional glass construction
- No special storage requirements
- Standard assembly techniques.
- UL 94 V-0 flammability rating
- **Absolutely the best cost performance available**

UL Information

File .... #E87492
Grade MC3D
Max Operating Temperature
Electrical .... 140°C
Mechanical .... 130°C
Flame Class 94V-0
Solder Limits
0.031" .... 525°F, 20 sec
0.062" .... 550°F, 30 sec

Electrical Properties			
	Frequency	@ 25°C	
Dielectric Constant	1.0 GHz	3.86 ± 0.08	
Dissipation Factor	1.0 GHz	0.019	
dB/inch Loss (S_{21} parameter from a 50 ohm 10 inch long transmission line)	100 MHZ	0.013	
	500 MHZ	0.048	
	1000 MHz	0.083	
	3000 MHz	0.233	
Surface Resistivity	C-96/35/90	Megohm	1×10^{10}
Volume Resistivity	C-96/35/90	Megohm-cm	1.5×10^{9}
Arc Resistance	A	seconds	60
Physical Properties			
Copper Peel Strength	A	Lb / inch width	6.0
	After 20 sec @500°F	Lb / inch width	6.0
Flex Strength Length	A	psi	56,000
Flex Strength Cross	A	psi	35,500
Water Absorption	D-24/23	%	0.22
Thermal Properties			
Glass Transition (Tg)	A	°C	120
Thermal Stress	@ 500 °F	seconds	40+
Thermal Conductivity	@ 121°C	W/m/K	0.327
Z-Axis CTE, RT → Tg	A	ppm/°C	88
Z-Axis CTE, Tg →260°C	A	ppm/°C	245
Dimensional Stability, Length	E-4/105, E-2/150	inch/inch	-0.00046
Dimensional Stability, Cross	E-4/105, E-2/150	inch/inch	-0.00072

- Typical Properties of 0.031" thick laminate, clad with 1 ounce copper. Properties of other thicknesses and copper weights may vary.
- All test methods are IPC-TM-650(Except Dk & Df)
- Tg measured by DMA

Rev D **TSR-152**

175 Commerce Road • Collierville, TN 38017 • 901-853-5070 • Fax 901-853-8664 • gilam@gilam.com

MC3D Medium Frequency Laminate

Typical Properties of MC3D Double Sided (0.062 inch)

Description

Grade MC3D is a composite laminate targeted for the rigid, double-sided market. Comprised of woven glass surface sheets on either side of a glass paper core, MC3D can be drilled cleanly with a minimum amount of drill wear. The unique resin system is compatible with conventional etchback and plated through hole chemistries. In addition, MC3D exhibits excellent electricals with a low and stable dielectric constant and dissipation factor. MC3D is an excellent material for frequency sensitive applications.

Features and Benefits

- Fabrication in conventional PWB (FR4) shops
- Standard FR4 drilling / cutting feeds and speeds.
- No special through hole treatments
- Conventional glass construction
- No special storage requirements
- Standard assembly techniques.
- UL 94 V-0 flammability rating
- **Absolutely the best cost performance available**

UL Information

File .... #E87492
Grade MC3D
Max Operating Temperature
Electrical .... 140°C
Mechanical......... 130°C
Flame Class....... 94V-0
Solder Limits
0.031"................ 525°F, 20 sec
0.062"................ 550°F, 30 sec

Electrical Properties			
	Frequency	@ 25°C	
Dielectric Constant	1.0 GHz	3.53 ± 0.08	
Dissipation Factor	1.0 GHz	0.017	
dB/inch Loss (S_{21} parameter from a 50 ohm 10 inch long transmission line	100 MHz	0.006	
	500 MHz	0.039	
	1000 MHz	0.075	
	3000 MHz	0.230	
Surface Resistivity	C-96/35/90	Megohm	1 X 10^{10}
Volume Resistivity	C-96/35/90	Megohm-cm	5 X 10^{8}
Arc Resistance	A	seconds	60
Physical Properties			
Copper Peel Strength	A	Lb / inch width	7.0
	After 20 sec @500°F	Lb / inch width	7.0
Flex Strength Length	A	psi	43,000
Flex Strength Cross	A	psi	26,500
Water Absorption	D-24/23	%	0.14
Thermal Properties			
Glass Transition (Tg)	A	°C	120
Thermal Stress	@ 500 °F	seconds	40+
Thermal Conductivity	@ 121°C	W/m/K	0.299
Z-Axis CTE, RT → Tg	A	ppm/°C	100
Z-Axis CTE, Tg → 260°C	A	ppm/°C	360
Dimensional Stability, Length	E-4/105, E-2/150	inch/inch	-0.00055
Dimensional Stability, Cross	E-4/105, E-2/150	inch/inch	-0.00090

- Typical Properties of 0.062" thick laminate, clad with 1 ounce copper. Properties of other thicknesses and copper weights may vary.
- All test methods are IPC-TM-650(Except Dk & Df)
- Tg measured by DMA

Rev L **TSR-131**

175 Commerce Road • Collierville, TN 38017 • 901-853-5070 • Fax 901-853-8664 • gilam@gilam.com

GML 1000 High Frequency Laminate
(Typical Properties of 0.020 ±0.002 inch Thickness)

Electrical Properties

Electrical Properties	Frequency	-55°C	25°C	80°C
Dielectric Constant Per IPC 650 2.5.5.5	2.5 GHz	3.05±0.05	3.05±0.05	3.05±0.05
	10.0 GHz	3.05±0.05	3.05±0.05	3.05±0.05
Dissipation Factor Per IPC 650 2.5.5.5	2.5 GHz	0.002	0.003	0.005
	10.0 GHz	0.003	0.005	0.006
dB/inch Loss (S_{21} parameter from a 50 ohm 10 inch long transmission line)	1.0 GHz		0.039	
	5.0 GHz		0.162	
	10.0 GHz		0.362	
Surface Resistivity	C-96/35/90		Megohm	$6X10^7$
Volume Resistivity	C-96/35/90		Megohm-cm	$8X10^9$
Moisture Insulation Resistance	20 cycles -2°C/90% RH to 65°C/95% RH		Megohm	$1X10^7$
Solvent Extract Conductivity	--		μg/cm²	0.39

Physical Properties

Physical Properties		Condition	Units	Value
Copper peel strength (1 oz copper)		Condition A	Lb /inch width	4.5
		After 20 sec @500°F	Lb /inch width	4.5
Flexural Strength	Length	Condition A	PSI	32,000
	Cross	Condition A	PSI	29,000
Flexural Modulus	Length	Condition A	MPSI	2.0
	Cross	Condition A	MPSI	1.8
Moisture Absorption		D-24/23	%	0.11

Thermal Properties

Thermal Properties		Condition	Units	Value
Glass Transition Temperature (DMA)		Condition A	°C	140
Thermal Stress		@500°F	seconds	40+
Z-Axis Expansion RT→T_g		Condition A	ppm/°C	60
Z-Axis Expansion T_g→260°C		Condition A	ppm/°C	450
X/Y Axis Expansion		Condition A	ppm/°C	40 / 40
Thermal Conductivity		@120°C	W/m/K	0.177
Dimensional Stability	Length	E-4/105 + E-2/150	inch/inch	-0.00220
	Cross	E-4/105 + E-2/150	inch/inch	-0.00192

Description

GIL Technologies new GML 1000 copper clad laminate has been specifically designed for the High Frequency Microstrip Antenna and Wireless Communications Market. GML 1000's dielectric constant (Dk) is low and stable when used over broad temperature and humidity operating ranges. The low insertion loss makes GML 1000 the most cost effective option when compared to PTFE and other recognized microwave laminates.

Features and Benefits

- Dk stable -55°C to 125°C
- Dk stable in humid and dry environments
- UL 94 V-0 Approved. File #E87492 (Oxygen Index >40%)
- No special through-hole treatments
- Fabrication and assembly in standard PWB operations
- Standard FR4 feeds & speeds for drilling & routing
- Excellent mechanical and electrical properties
- **Absolutely the BEST cost performance available**

- Typical properties of 020" laminate clad with 1 ounce copper. Properties of other thicknesses and copper weights may vary.

Rev D **TSR-156**

175 Commerce Road • Collierville, TN 38017 • 901-853-5070 • Fax 901-853-8664 • gilam@gilam.com

GML 1000 High Frequency Laminate
(Typical Properties of 0.030 ±0.002 inch Thickness)

Electrical Properties	Frequency	-55°C	25°C	80°C
Dielectric Constant	2.5 GHz	3.20±0.05	3.20±0.05	3.20±0.05
per IPC-TM-650 2.5.5.5	10.0 GHz	3.20±0.05	3.20±0.05	3.20±0.05
Dissipation Factor	2.5 GHz	0.002	0.003	0.004
per IPC-TM-650 2.5.5.5	10.0 GHz	0.004	0.004	0.005
dB/inch Loss	1.0 GHz		0.029	
(S_{21} parameter from a 50 ohm	5.0 GHz		0.125	
10 inch long transmission line)	10.0 GHz		0.277	
Surface Resistivity	C-96/35/90		Megohm	$5X10^7$
Volume Resistivity	C-96/35/90		Megohm-cm	$8X10^9$
Moisture Insulation Resistance	20 cycles -2°C/90% RH to 65°C/95% RH		Megohm	$1X10^7$
Solvent Extract Conductivity	--		µg/cm²	0.53

Physical Properties				
Copper peel strength		Condition A	Lb /inch width	5.0
(1 oz copper)		After 20 sec @500°F	Lb /inch width	5.0
Flexural Strength	Length	Condition A	PSI	43,500
	Cross	Condition A	PSI	38,000
Flexural Modulus	Length	Condition A	MPSI	2.3
	Cross	Condition A	MPSI	2.1
Moisture Absorption		D-24/23	%	0.06

Thermal Properties				
Glass Transition Temperature (DMA)		Condition A	°C	135
Thermal Stress		@500°F	seconds	40+
Z-Axis Expansion RT→T_g		Condition A	ppm/°C	70
Z-Axis Expansion T_g→260°C		Condition A	ppm/°C	400
X/Y Axis Expansion		Condition A	ppm/°C	32 / 32
Thermal Conductivity		@120°C	W/m/K	0.228
Dimensional Stability	Length	E-4/105 + E-2/150	inch/inch	-0.00066
	Cross	E-4/105 + E-2/150	inch/inch	-0.00075

Description

GIL Technologies new GML 1000 copper clad laminate has been specifically designed for the High Frequency Microstrip Antenna and Wireless Communications Market. GML 1000's dielectric constant (Dk) is low and stable when used over broad temperature and humidity operating ranges. The low insertion loss makes GML 1000 the most cost effective option when compared to PTFE and other recognized microwave laminates.

Features and Benefits

- Dk stable -55°C to 125°C
- Dk stable in humid and dry environments
- UL 94 V-0 Approved. File #E87492 (Oxygen Index >40%)
- No special through-hole treatments
- Fabrication and assembly in standard PWB operations
- Standard FR4 feeds & speeds for drilling & routing
- Excellent mechanical and electrical properties
- **Absolutely the BEST cost performance available**

- Typical properties of 030" laminate clad with 1 ounce copper. Properties of other thicknesses and copper weights may vary.

Rev D **TSR-154**

175 Commerce Road • Collierville, TN 38017 • 901-853-5070 • Fax 901-853-8664 • gilam@gilam.com

GML 1000 High Frequency Laminate
(Typical Properties of 0.060 ±0.003 inch Thickness)

Electrical Properties

Electrical Properties	Frequency	-55°C	25°C	80°C
Dielectric Constant	2.5 GHz	3.05±.05	3.05±.05	3.05±.05
per IPC-TM-650 2.5.5.5	10.0 GHz	3.05±.05	3.05±.05	3.05±.05
Dissipation Factor	2.5 GHz	0.002	0.003	0.004
per IPC-TM-650 2.5.5.5	10.0 GHz	0.003	0.004	0.005
dB/inch Loss	1.0 GHz		0.023	
(S_{21} parameter from a 50 ohm	5.0 GHz		0.123	
10 inch long transmission line)	10.0 GHz		0.295	
Surface Resistivity	C-96/35/90		Megohm	2 X 10^8
Volume Resistivity	C-96/35/90		Megohm-cm	8 X 10^8
Moisture Insulation Resistance	20 cycles -2°C/90% RH to 65°C/95% RH		Megohm	1 X 10^7
Solvent Extract Conductivity	--		μg/cm²	0.49

Physical Properties

Physical Properties				
Copper peel strength		Condition A	Lb /inch width	5.0
(1 oz copper)		After 20 sec @500°F	Lb /inch width	5.0
Flexural Strength	Length	Condition A	PSI	25,000
	Cross	Condition A	PSI	22,500
Flexural Modulus	Length	Condition A	MPSI	1.6
	Cross	Condition A	MPSI	1.5
Moisture Absorption		D-24/23	%	0.02

Thermal Properties

Thermal Properties				
Glass Transition Temperature (DMA)		Condition A	°C	135
Thermal Stress		@500°F	seconds	40+
Z-Axis Expansion RT→T_g		Condition A	ppm/°C	80
Z-Axis Expansion T_g→260°C		Condition A	ppm/°C	410
X/Y Axis Expansion		Condition A	ppm/°C	40 / 34
Thermal Conductivity		@120°C	W/m/K	0.170
Dimensional	Length	E-4/105 + E-2/150	inch/inch	-0.00070
Stability	Cross	E-4/105 + E-2/150	inch/inch	-0.00070

Description

GIL Technologies new GML 1000 copper clad laminate has been specifically designed for the High Frequency Microstrip Antenna and Wireless Communications Market. GML 1000's dielectric constant (Dk) is low and stable when used over broad temperature and humidity operating ranges. The low insertion loss makes GML 1000 the most cost effective option when compared to PTFE and other recognized microwave laminates.

Features and Benefits

- Dk stable -55°C to 125°C
- Dk stable in humid and dry environments
- UL 94 V-0 Approved. File #E87492 (Oxygen Index >40%)
- No special through-hole treatments
- Fabrication and assembly in standard PWB operations
- Standard FR4 feeds & speeds for drilling & routing
- Excellent mechanical and electrical properties
- **Absolutely the BEST cost performance available**

- Typical properties of 060" laminate clad with 1 ounce copper. Properties of other thicknesses and copper weights may vary.

Rev D **TSR-153**

175 Commerce Road • Collierville, TN 38017 • 901-853-5070 • Fax 901-853-8664 • gilam@gilam.com

DATA

RO 1.3000

RO3000™ Series
High Frequency Circuit Materials

RO3000 High Frequency Circuit Materials are ceramic filled PTFE composites intended for use in commercial microwave and RF applications. This family of products was designed to offer exceptional electrical and mechanical stability at competitive prices.

RO3000 series is a line of PTFE materials with mechanical properties that are the same regardless of dielectric constant. This allows the designer to develop multi-layer board designs that use different dielectric constant materials for individual layers, without encountering warpage or reliability problems.

The dielectric constant versus temperature of RO3000 series materials is very stable (Charts 1 and 2). These materials exhibit a thermal coefficient of expansion (CTE) in the X and Y axis of 17 ppm/°C. This expansion coefficient is matched to that of copper which allows the material to exhibit excellent dimensional stability, with typical etch shrinkage (after etch and bake) of less then 0.5 mils per inch. The Z-axis CTE is 24 ppm/ °C, which provides exceptional plated through-hole reliability, even in severe thermal environments.

RO3000 series laminates can be fabricated into printed circuit boards using standard PTFE circuit board processing techniques with minor modifications as described in the application note "Fabrication Guidelines for RO3000 Series High Frequency Circuit Materials."

Cladding options include ½ to 2 oz./ft^2 (17 to 68 μm thick) electrodeposited copper.

RO3000 laminates are manufactured under an ISO 9002 certified system.

RO3000™ Laminate Product Information:

PROPERTY	TYPICAL VALUE			DIRECTION	UNITS	CONDITION	TEST METHOD
	RO3003	RO3006	RO3010				
Dielectric Constant ε_r	3.0±0.04[(2)]	6.15±0.15	10.2±0.30	Z	—	10 GHz 23°C	IPC-TM-650, 2.5.5.5
Thermal Coefficient of ε_r	13	-169	-295	Z	ppm/°C	10 GHz 0-100°C	IPC-TM-650, 2.5.5.5
Dissipation Factor	0.0013	0.0025	0.0035	Z	—	10 GHz 23°C	IPC-TM-650 2.5.5.5
Dimensional Stability	0.5	0.5	0.5	X,Y	mm/m	A	ASTM D257
Volume Resistivity	10^6	10^3	10^3	Z	Mohm-cm	A	IPC 2.5.17.1
Surface Resistivity	10^7	10^3	10^3	A	Mohm	A	IPC 2.5.17.1
Tensile Modulus	2070 (300)	2070 (300)	2070 (300)	X,Y	MPa (kpsi)	23°C	ASTM D638
Water Absorption	<0.1	<0.1	<0.1	—	%	D24/23	IPC-TM-650 2.6.2.1
Specific Gravity	2.1	2.6	3.0	—	—	23°C	ASTM D792
Copper Peel Strength	3.1 (17.6)	2.1 (12.2)	2.4 (13.4)	—	N/mm (lb/in)	After Solder Float	IPC-TM-650 2.4.8
Specific Heat	0.93 (0.22)	0.93 (0.22)	0.93 (0.22)	—	J/g/K (BTU/lb/°F)	—	Calculated
Thermal Conductivity	0.50	0.61	0.66	—	W/m/K	100°C	ASTM C518
Coefficient of Thermal Expansion	17 24	17 24	17 24	X,Y Z	ppm/°C	-55 to 288°C	ASTM D3386-94
Color	Tan	Tan	Off White	—	—	—	—
Density	2.1	2.6	3.0	—	gm/cm^3	—	—
UL Flammability Rating	94V-O	94V-O	94V-O	—	—	—	—

(1) References: Internal T.R.'s 1430, 2224, 2854. Tests at 23°C unless otherwise noted. Typical values should not be used for specification limits.

(2) The nominal dielectric constant of an 0.060" thick RO3003 laminate as measured by the IPC-TM-650, 2.5.5.5 will be 3.02, due to the elimination of biasing caused by air gaps in the test fixture. For further information refer to Rogers T.R. 5242.

The data in Charts 1, 2 and 3 was produced using a modified IPC TM-650, 2.5.5.5 method. For additional information request Rogers T.R. 5156 and T.M. 4924.

Revised date: 10/98

The data in Chart 1 demonstrates the excellent stability of dielectric constant over temperature for RO3003™ laminates, including the elimination of the step change in dielectric constant which occurs near room temperature with PTFE glass materials.

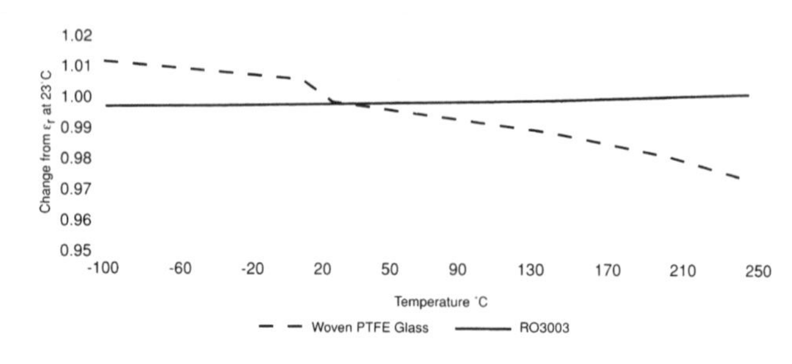

The data in Chart 2 shows the change in dielectric constant vs temperature for RO3006 and 3010 laminates. These materials exhibit significant improvement in temperature stability of dielectric constant when compared to other high dielectric constant PTFE laminates.

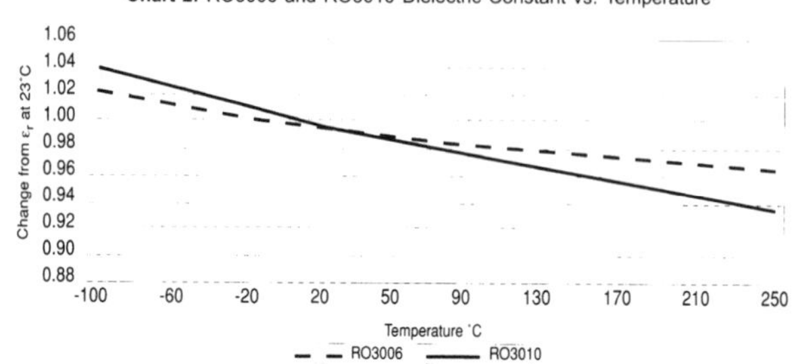

Chart 3 demonstrates the stability of dielectric constant for RO3000 series products over frequency. This stability simplifies the design of broad band components as well as allowing the materials to be used in a wide range of applications over a very broad range of frequencies.

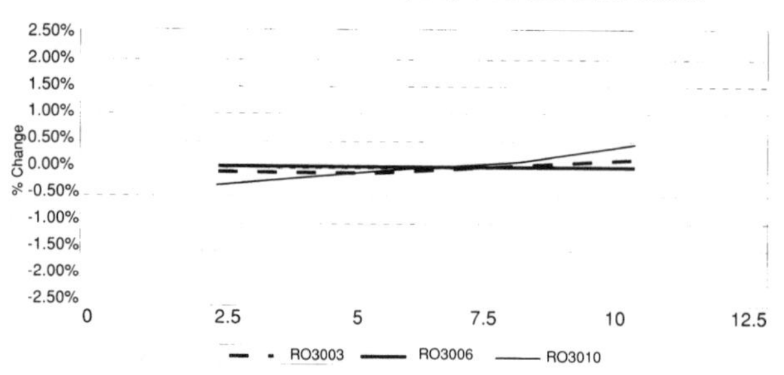

Ordering Information:

Available Configurations:
Thicknesses and Tolerance

RO3003	RO3006	RO3010
0.005" ± 0.0005" (0.13 ± 0.01 mm) 0.010" ± 0.0007" (0.25 ± 0.02 mm) 0.020" ± 0.001" (0.50 ± 0.03 mm) 0.030" ± 0.0015" (0.75 ± 0.04 mm) 0.060" ± 0.003" (1.52 ± 0.08 mm)	0.005"± 0.0005" (0.13 ± 0.01 mm) 0.010"± 0.0007" (0.25 ± 0.02 mm) 0.025"± 0.001" (0.64 ± 0.03 mm) 0.050"± 0.002" (1.27 ± 0.05 mm)	0.005"± 0.0005" (0.13 ± 0.01 mm) 0.010"± 0.0007" (0.25 ± 0.02 mm) 0.025"± 0.001" (0.64 ± 0.03 mm) 0.050"± 0.002" (1.27 ± 0.05 mm)

Standard Claddings
1/2 oz (17 µm) electrodeposited foil
1 oz (35 µm) electrodeposited foil
2 oz (70 µm) electrodeposited foil

Standard Panel Sizes
The standard panel size is 24" X 18" (620 X 457 mm), and 12 X 18" (305 X 457 mm).

These products may require an export license issued by the United States Department of Commerce for export of these materials from the United States or Canada.

The information and guidelines contained in this document are intended to assist you in designing with RO3000 High Frequency Circuit Materials. They do not create any warranties express or implied, including any warranty of merchantability or fitness for a particular purpose. Results may vary as conditions and equipment vary. The user should determine the suitability of Rogers materials for each application. Values are averages and not guaranteed.

ROGERS
SINCE 1832

Rogers Corporation
Microwave Materials Division
100 S. Roosevelt Avenue
Chandler, AZ 85226-3415
602 961-1382 FAX: 602 961-4533
Toll Free: 877 643-7701
http://www.rogers-corp.com/mwu/
ISO 9002 CERTIFIED

In Japan:
Rogers Japan Inc.,7th Floor, ST Bldg, 2-26-9 Nishi-nippori,
Arakawa-ku, Tokyo 116 Japan
03-3807-6430 FAX: 03-3807-6319

In Hong Kong:
Rogers Southeast Asia, 21st Floor, Unit 2
118 Connaught Road West, Sheung Wan Hong Kong
852-2549-7806 FAX: 852-2549-8615

In Europe:
Rogers N.V., Afrikalaan 188, B-9000 Gent, Belgium
32-9-2353611 FAX: 32-9-2353658

In Taiwan:
Rogers Taiwan Inc., 15F, No. 343, Chung-Ho Road
Yung-Ho City, Taipei Hsien, Taiwan R.O.C.
886-2-86609056 FAX: 886-2-86609057

Printed in U.S.A. Revised 10/98 1308-108-5.0-ON

RO4000® Series
High Frequency Circuit Materials

RO4000® Series High Frequency Circuit Materials are glass reinforced hydrocarbon/ceramic laminates (**Not PTFE**) designed for performance sensitive, high volume commercial applications.

RO4000 laminates are designed to offer superior high frequency performance and low cost circuit fabrication. The result is a low loss material which can be fabricated using standard epoxy/glass (FR4) processes offered at competitive prices.

The selection of laminates typically available to designers is significantly reduced once operational frequencies increase to 500 MHz and above. RO4000 material possesses the properties needed by designers of RF Microwave circuits. Stable electrical properties over environmental conditions allow for repeatable design of filters, matching networks and controlled impedance transmission lines. Low dielectric loss allows RO4000 series material to be used in many applications where higher operating frequencies limit the use of conventional circuit board laminates. The temperature coefficient of dielectric constant is among the lowest of any circuit board material (Chart 1), making it ideal for temperature sensitive applications. RO4000 materials exhibit a stable dielectric constant over a broad frequency range (Chart 2). This makes it an ideal substrate for broadband applications.

RO4000 material's thermal coefficient of expansion (CTE) provides several key benefits to the circuit designer. The expansion coefficient of RO4000 material is similar to that of copper which allows the material to exhibit excellent dimensional stability, a property needed for mixed dielectric multilayer board constructions. The Z-axis CTE provides reliable plated through-hole quality, even in severe thermal shock applications. RO4000 series material has a Tg of >280°C so its expansion characteristics remain stable over the entire range of circuit processing temperatures.

RO4000 series laminates can easily be fabricated into printed circuit boards using standard FR4 circuit board processing techniques. Unlike PTFE based high performance materials, RO4000 series laminates do not require specialized processes such as sodium etch. This material is a rigid laminate that is capable of being processed by automated handling systems and scrubbing equipment used for copper surface preparation.

RO 1.4000
Page 2 of 4

RO4000® Series Laminate Product Information:

PROPERTY	TEST METHOD	CONDITION	UNITS	DIRECTION	RO4003 TYPICAL VALUES	RO4350 TYPICAL VALUES
Dielectric Constant ε_r	IPC-TM-650 2.5.5.5	10 GHz/23°C	-	Z	3.38 ± 0.05	3.48 ± 0.05
Dissipation Factor	IPC-TM-650 2.5.5.5	10 GHz/23°C	-	Z	0.0027	0.0040
Thermal Coefficient of ε_r	IPC-TM-650 2.5.5.5	-100°C to 250°C	ppm/°C	Z	+40	+50
Volume Resistivity	IPC-TM-650 2.5.17.1	COND A	MΩ	–	1.7×10^{10}	1.2×10^{10}
Surface Resistivity	IPC-TM-650 2.5.17.1	COND A	MΩ	-	4.2×10^{9}	5.7×10^{9}
Electrical Strength	IPC-TM-650 2.5.6.2	0.51mm (0.020")	KV/mm (V/mil)	Z	25.6 (650)	31.5 (800)
Tensile Modulus	ASTM D638	RT	MPa (kpsi)	Y	26,889 (3900)	11,473 (1664)
Tensile Strength	ASTM D638	RT	MPa (kpsi)	Y	141 (20.4)	175 (25.4)
Flexural Strength	IPC-TM-650 2.4.4	-	MPa (kpsi)	-	276 (40)	255 (37)
Dimensional Stability	IPC-TM-650 2.24	After Etch +E2/150	mm/m (mils/inch)	X,Y	<0.3	<0.5
Coefficient of Thermal Expansion	IPC-TM-650 2.1.4.1	-55 to 288°C	ppm/°C	X Y Z	11 14 46	14 16 50
Tg	TMA	-	°C	-	>280	>280
Thermal Conductivity	ASTM F433	100°C	W/m/°K	-	0.64	0.62
Specific Gravity	ASTM D792	23°C	-	-	1.79	1.86
Water Absorption	ASTM D570	48 hrs. immersion 0.60" sample Temperature 50°C	%	-	0.06	0.06
Copper Peel Strength	IPC-TM-650 2.48	after solder float	N/mm (pli)	-	1.05 (6.0)	0.88 (5.0)
Flammability Rating	UL 94V-0	-	-	-	No	Yes

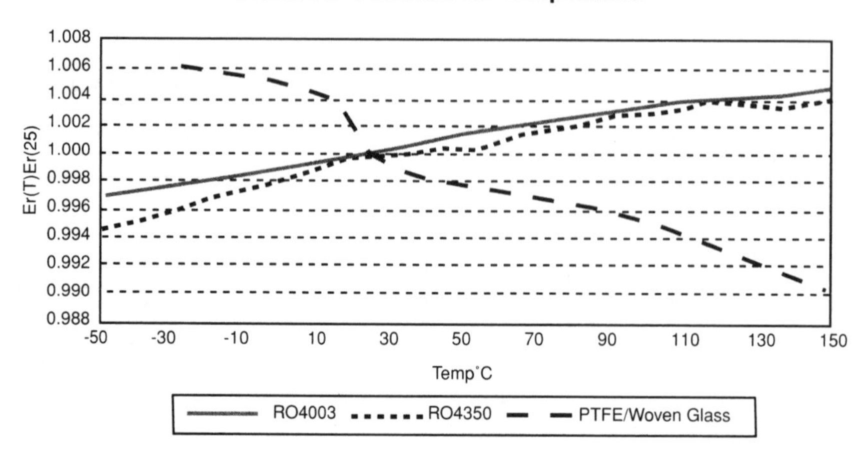
Chart 1: RO4000 Series Materials
Dielectric Constant vs. Temperature
1.008
1.006
1.004
1.002
1.000
0.998
0.996
0.994
0.992
0.990
0.988
Er(T)Er(25)
-50
-30
-10
10
30
50
70
90
110
130
150
Temp°C
RO4003
RO4350
PTFE/Woven Glass

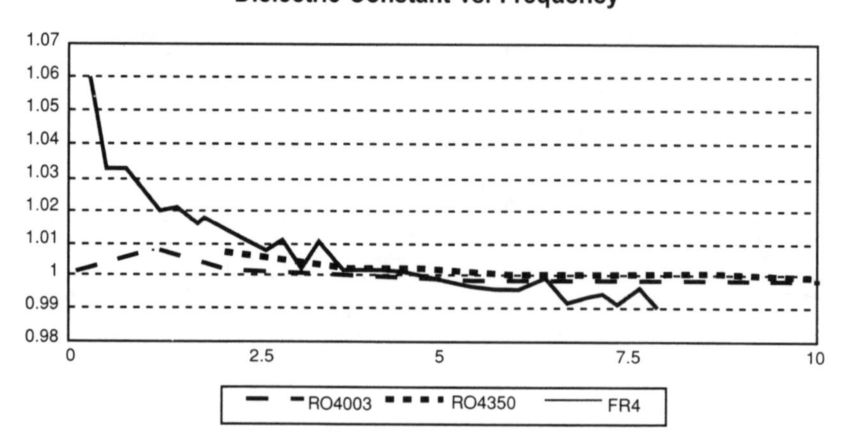
Chart 2: RO4000 Series Materials
Dielectric Constant vs. Frequency
1.07
1.06
1.05
1.04
1.03
1.02
1.01
1
0.99
0.98
0
2.5
5
7.5
10
RO4003
RO4350
FR4

Chart 3: Microstrip Insertion Loss
(0.030" Dielectric Thickness)

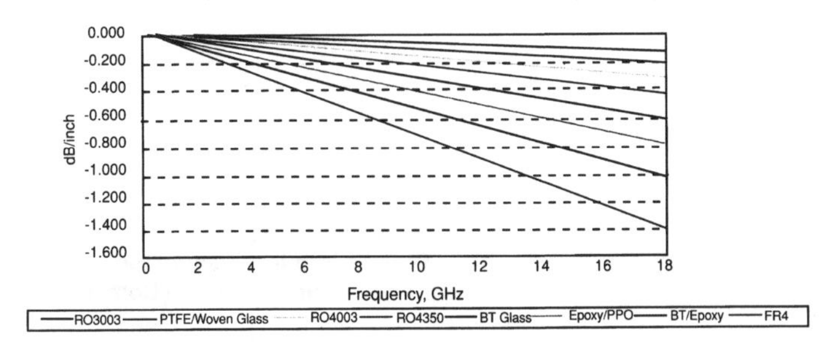

Ordering Information:

Standard Thicknesses and Tolerances:

RO4003 Materials:

0.0080 ± 0.0010 (0.20 ± 0.03 mm)
0.0200 ± 0.0015 (0.51 ± 0.04 mm)
0.0320 ± 0.0020 (0.81 ± 0.05 mm)
0.0600 ± 0.0040 (1.52 ± 0.10 mm)

RO4350 Materials:

0.0066 ± 0.0007 (0.17 ± 0.02 mm)
0.0100 ± 0.0010 (0.25 ± 0.03 mm)
0.0200 ± 0.0015 (0.51 ± 0.04 mm)
0.0300 ± 0.0020 (0.76 ± 0.05 mm)
0.0600 ± 0.0040 (1.52 ± 0.10 mm)

Standard Claddings:

½ ounce (17 μm) electrodeposited copper
1 ounce (35 μm) electrodeposited copper

Standard Panels Sizes:

The standard panel sizes are 24 x 18" (620 x 457 mm) and 12 x 18" (305 x 457 mm).

Information on other thicknesses, claddings and panels sizes available call your Customer Service Representative at 602 961-1382 or Fax: 602 961-4533.

ROGERS
SINCE 1832

Rogers Corporation
Microwave Materials Division
100 S. Roosevelt Avenue
Chandler, AZ 85226-3415
602 961-1382 FAX: 602 961-4533
http://www.rogers-corp.com/mwu/
ISO 9002 CERTIFIED

In Japan:
Rogers Japan Inc.
7th Floor, ST Bldg
2-26-9 Nishi-nippori
Arakawa-ku
Tokyo 116 Japan
03-3807-6430
FAX: 03-3807-6319

In Hong Kong:
Rogers Southeast Asia, Inc.
21st Floor, Unit 2
118 Connaught Road West
Sheung Wan, Hong Kong
Tel: 852-2549-7806,
FAX: 852-2549-8615

In Europe:
Rogers N.V.
Afrikalaan 188
B-9000 Gent
BELGIUM
Tel: 32-9-235-3611
Fax: 32-9-235-3658

Printed in U.S.A.

ISSUE DATE: 6/98 1282-068-2.5-ON

RT/duroid® 5870
Glass Microfiber Reinforced
Polytetrafluoroethylene Composite

RT/duroid® 5870 glass microfiber reinforced PTFE composite is designed for exacting stripline and microstrip circuit applications.

Glass reinforcing microfibers are randomly oriented to maximize benefits of fiber reinforcement in the directions most valuable to circuit producers and in the final circuit application.

The dielectric constant of RT/duroid 5870 laminates is uniform from panel to panel and is constant over a wide frequency range. Its low dissipation factor extends the usefulness of RT/duroid 5870 to X-band and above.

RT/duroid 5870 laminate is easily cut, sheared and machined to shape. It has excellent dimensional stability and is resistant to all solvents and reagents hot or cold, normally used in etching printed circuits or in plating edges and holes.

Normally supplied as a laminate with electrodeposited copper of 1/4 to 2 ounces/ft.^{2} on both sides, RT/duroid 5870 composites can also be clad with rolled copper foil for more critical electrical applications. Cladding with aluminum, copper and brass plate may also be specified.

When requested copper clad, RT/duroid 5870 composite can be certified to MIL-S-13949H Type GRN or Type GPN microwave material specifications.

When ordering RT/duroid 5870 laminates, it is important to specify dielectric thickness, tolerance, rolled or electrodeposited copper foil and weight of copper foil required.

(See reverse for product data)

ROGERS **RT/duroid 5870**

PROPERTY	TEST METHOD	CONDITION	UNITS[1]	DIRECTION	TYPICAL VALUE [2]		
Dieletric Constant, ε_r	1 MHz, MIL-S-13949H	C24/23/50	--	Z	2.35		
	10 GHz, IPC-TM-650, 2.5.5.5	C24/23/50	--	Z	2.33 ± 002 spec.		
Dissipation factor, tan δ	1 MHz, MIL-S-13949H	C24/23/50	--	Z	0.005		
	10 GHz, IPC-TM-650, 2.5.5.5	C24/23/50	--	Z	0.0012		
Volume resistivity	ASTM D257	C96/35/90	Mohm cm	Z	2×10^7		
Surface resistivity	ASTM D257	C96/25/90	Mohm	Z	2×10^8		
					Test at 23°C	Test at 100°C	
Tensile Modulus	ASTM D638	A	MPa(kpsi)	X	1300(189)	490(71)	
				Y	1280(185)	430(63)	
ultimate stress			MPa(kpsi)	X	50 (7.3)	34(4.8)	
				Y	42(6.1)	34(4.8)	
ultimate strain			%	X	9.8	8.7	
				Y	9.8	8.6	
Compressive modulus	ASTM D695	A	MPa(kpsi)	X	1210(176)	680(99)	
				Y	1360(198)	860(125)	
				Z	830(120)	520(76)	
ultimate stress			MPa(kpsi)	X	30(4.4)	23(3.4)	
				Y	37(5.3)	25(3.7)	
				Z	54(7.8)	37(5.3)	
ultimate strain			%	X	4.0	4.3	
				Y	3.3	3.3	
				Z	8.7	8.	
Water absorption	ASTM D570	D24/23					
Thickness = 0.9mm(0.031in.)				mg(%)		0.9(0.02)	
Thickness = 1.6mm(0.062in.)				mg(%)		1.3(0.015)	
Specific gravity	ASTM D792				2.2		
Heat distortion temperature	ASTM D648	1.82MPa(264psi)	°C(°F)	X,Y	>260(>500)		
Specific heat	Calculated		J/g/K(BTU)/lb/°F)		0.96(0.23)		
Thermal conductivity	Rogers TR2721		W/m/K(BTU in/ft²/hr°F)	Z	0.26(1.8)		
Thermal expansion	ASTM D3386		mm/m	→	X	Y	Z
	(10 K/min.)	-100°C			-5.0	-5.5	-11.6
(Values given are total change from a base temperature of 35°C)		15			-0.6	-0.9	- 4.0
		25			-0.3	-0.4	- 2.6
		75			0.7	0.9	7.5
		150			1.8	2.2	22.0
		250			3.4	4.0	58.9

[1] SI units given first with other frequently used units in Parentheses.
[2] References: Internal TR's 1430, 2224, 2854. Tests were at 230C unless otherwise noted. Typical values should not be used for specification limits.

The information contained in this document is intended to assist you in designing with RT/duroid microwave laminates. They are not intended to and do not create any warranties, express or implied, including any warranty of merchantability or fitness for a particular application. Failure to follow these guidelines may negate any warranties that otherwise exist. Results may vary as conditions and equipment may vary. The user should determine the suitability of Rogers materials for each specific application.

These product may required a validated export license issued by the U.S. Department of Commerce for export of these materials from the United States or Canada.

RT/duroid is a registered trademark of Rogers Corporation for its microwave laminates.

ROGERS

Rogers Corporation
Microwave and Circuit Materials Division, 100 S. Roosevelt Chandler, AZ 85226 602 961-1382 FAX: 602 961-4533

Revised 4/93
Supersedes 5/91
6735-043-10.0-AH

RT 1.5880

RT/duroid®

RT / duroid® 5880
Glass Microfiber Reinforced Polytetrafluoroethylene Composite

RT/duroid® 5880 glass microfiber reinforced PTFE composite is designed for exacting stripline and microstrip circuit applications.

Glass reinforcing microfibers are randomly oriented to maximize benefits of fiber reinforcement in the directions most valuable to circuit producers and in the final circuit application.

The dielectric constant of RT/duroid 5880 laminates is uniform from panel to panel and is constant over a wide frequency range. Its low dissipation factor extends the usefulness of RT/duroid 5880 to Ku-band and above.

RT/duroid 5880 laminate is easily cut, sheared and machined to shape. It has excellent dimensional stability and is resistant to all solvents and reagents, hot or cold, normally used in etching printed circuits or in plating edges and holes.

Normally supplied as a laminate with electrodeposited copper of 1/4 to 2 ounces/ ft.^{2} on both sides, RT/duroid 5880 composites can also be clad with rolled copper foil for more critical electrical applications. Cladding with aluminum, copper or brass plate may also be specified.

When requested copper-clad, RT/duroid 5880 composite can be certified to MIL-S-13949 Type GRN or Type GPN microwave material specifications.

When ordering RT/duroid 5880 laminates, it is important to specify dielectric thickness, tolerance, rolled or electrodeposited copper foil, and weight of copper foil required.

(See reverse for product data)

ROGERS

RT/duroid 5880 Microwave Laminate

PROPERTY	TEST METHOD	CONDITION	UNITS[1]	DIRECTION	TYPICAL VALUE[2]		
Dielectric constant, ε_r	IMHz,IPC-TM-650, 2.5.5.3	C 24/23/50	—	Z	2.20		
	10GHz, IPC-TM-650, 2.5.5.5	C 24/23/50			2.20± 0.02 (± 0.01 available)		
Dissipation factor, tan δ	IMHz, IPC-TM-650-2.5.5.3	C24/23/50	—	Z	0.0004		
	10GHz, IPC-TM-650, 2.5.5.5	C24/23/50	—	Z	0.0009		
Volume resistivity	IPC-TM-650,2.5.17.1	C96/35/90	Mohm cm	Z	2×10^7		
Surface resistivity	IPC-TM -650,2.5.17.1	C96/35/90	Mohm	X,Y	3×10^8		
					Test at 23°C	Test at 100°C	
Tensile modulus	ASTM D638	A	MPa (kpsi)	X	1070 (156)	450 (65)	
				Y	860 (125)	380 (55)	
ultimate stress			MPa (kpsi)	X	29 (4.2)	20 (2.9)	
				Y	27 (3.9)	18 (2.6)	
ultimate strain			%	X	6.0	7.2	
				Y	4.9	5.8	
Compressive modulus	AST D695	A	MPa (kpsi)	X	710 (103)	500 (73)	
				Y	710 (103)	500 (75)	
				Z	940 (136)	670 (97)	
ultimate stress			MPa (kpsi)	X	27 (3.9)	22 (3.2)	
				Y	28 (4.0)	21 (3.1)	
				Z	52 (7.5)	43 (6.3)	
ultimate strain			%	X	8.5	8.4	
				Y	7.7	7.8	
				Z	12.5	17.6	
Deformation under load	ASTM D621				Test at 150°C		
		24 hr/14MPa(2 kpsi)	%	Z	1.0		
Water absorption	ASTM D570	D 24/23					
Thickness = 0.8mm (0.031 in.)			mg (%)		0.9 (0.02)		
Thickness = 1.6mm (0.062 in.)			mg (%)		1.3 (0.015)		
Specific gravity	ASTM D792		—		2.2		
Heat distortion temperature	ASTM D648	1.82MPa (264 psi)	°C(°F)	X, Y	>260 (>500)		
Specific heat	Calculated		J/g/K(BTU/lb/°F)		0.96 (0.23)		
Thermal conductivity	Rogers TR2721		W/m/K(BTU in/ft²/hr/°F)	Z	0.26 (1.8)		
Thermal expansion	ASTM D3386 (10 K/min.)		mm/m	→	**X**	**Y**	**Z**
		-100°C			-6.1	-8.7	-18.7
		15			-0.9	-1.8	-6.9
(Values given are total change from a base temperature of 35°C)		25			-0.5	-0 9	- 4.5
		75			1.1	1.5	8.7
		150			2.3	3.2	28.3
		250			3.8	5.5	69.5

[1] SI units given first with other frequently used units in parentheses.
[2] References: Internal TR's 1430, 2224, 2854. Tests were at 23°C unless otherwise noted. Typical values should not be used for specification limits.

The above information is not intended to and does not create any warranties, express or implied, including any warranty of merchantability or fitness for a particular purpose. Use of **RT/duroid** microwave laminate in your particular application may yield different results.

These products may require a validated export license issued by the U.S. Department of Commerce for export of these materials from the United States or Canada.

RT / duroid is a registered trademark of Rogers Corporation for its microwave laminates.

ROGERS
Rogers Corporation
Microwave and Circuit Materials Division, 100 S. Roosevelt Avenue, Chandler, AZ 85226, 602 961-1382, Fax: 602 961-4533

Revised 8/94
Supersedes 8/92
7306-094-5.0-AP

Printed in U.S.A.

RT6010LM

RT/duroid® 6010LM
Ceramic Polytetrafluoroethylene (PTFE) Composite

RT/duroid 6010LM microwave laminate is a ceramic - PTFE composite designed for microwave circuit applications requiring a high dielectric constant. This product's properties have been carefully tailored to achieve **L**ow **M**oisture absorption (typically 0.05%), dramatically reducing the effects of moisture on electrical loss. **RT/duroid** 6010LM is available with a dielectric constant of 10.2 ± 0.25.

RT/duroid 6010LM microwave laminate offers exceptional thickness control (typically < 4%), low Z axis expansion for plated thru-hole reliability, low levels of surface and subsurface contaminants and nearly isotropic electrical properties.

Laminates are supplied clad both sides with 1/8 to 2 oz./ft.^{2} electrodeposited (ED) copper foil and 1/2 to 2 oz./ft.^{2} rolled (R) copper foil. A wide variety of thick metal plates including aluminum, brass and copper may be specified. Peel strength after solder float for 1 oz./ft^2 ED copper foil is typically 10 lbs./in.

The standard available thicknesses are 0 .010", 0.025", 0.050", 0.075" and 0.100" with the material generally available from 0.005" to 0.750" in multiples of 0.005". The standard available sheet sizes are 10"x10", 10"x20" and 20"x20" with up to 22"x22" guaranteed electrically good. Material outside of the standard panel size (20"x20") may be requested as guaranteed mechanically good at no extra charge for tooling holes.

It is important to specify dielectric thickness, thickness tolerance, foil type, foil weight and sheet size when ordering **RT/duroid** 6010LM laminates for example, 6010LM 0.025 ±.001 1E/1E 10x10.

(See reverse for product data)

ROGERS

RT 1.6010LM

RT/duroid® 6010LM

PROPERTY	TEST METHOD	CONDITION	UNITS[1]	DIRECTION	TYPICAL VALUE[2]
Deilectric constant ε_r	Adapted IPC-TM-650, 2.5.5.5	A	--	Z	10.2 ±0.25 [3]
Thermal Coefficient of ε_r	Adapted IPC-TM-650, 2.5.5.5	-50 to 170°C	ppm/°C	Z	-370
Dissipation factor, tan δ	Adapted IPC-TM-650, 2.5.5.5	A	--	Z	0.0028 max., spec.
Tensile modulus	ASTM D638 (0.1/min. strain rate)	A	MPa(kpsi)	X	931(135)
				Y	559 (81)
ultimate stress		A	MPa(kpsi)	X	17(2.4)
				Y	13(1.9)
ultimate strain			%	X	9 to 15
				Y	7 to 14
Compression modulus	ASTM D695 (0.05/min. strain rate)	A	MPa(kpsi)	Z	2144(311)
ultimate stress			MPa(kpsi)	Z	47(6.9)
ultimate strain			%	Z	25
Flexural modulus	ASTM D790	A	MPa(kpsi)	X	4364(633)
				Y	3751(544)
ultimate stress			MPa(kpsi)	X	36(5.2)
				Y	32(4.4)
Deformation under load	ASTM D621	24 hr/50°C/7MPa	%	Z	0.26
		24 hr/150°C/7MPa	%	Z	1.37
Water absorption	IPC-TM-650, 2.6.2.1 ASTM D570	24 hr/23°C, 0.025"(.635 mm) thick	%		0.05 typical 0.2 maximum
Specific gravity	ASTM D792	--	--		2.9
Specific heat	Calculated		J/g/K(BTU/lb/°F)		1.00(0.239)
Thermal conductivity	Rogers TR 2721	23 to 100°C	W/m/K (BTU in/ft²/hr/°F)	Z	0.41 (2.87)
Thermal expansion	ASTM D3386 (5 K/min.)		mm/m		X Y Z
		-100°C			-2.8 -3.0 -3.4
		-50°C			-2.0 -2.1 -2.6
	(Values given are total change from a base temperature of 35°C)	10°C			-0.8 -0.8 -1.1
		75°C			1.0 1.0 0.7
		150°C			2.2 2.2 1.7
		250°C			3.7 3.8 4.3
		315°C			5.0 5.1 8.4

[1] S1 units given first, with other frequently used units in parentheses.
[2] References: APR 4022:44; DJS 4019:27-32; Internal TR 2610. Tests were at 23°C unless otherwise noted. Typical values should not be used for specifiction limits
[3] Based on testing a 0.025 thick sheet clad with 1 oz electrodeposited foil.

The above information is not intended to and does not create any warranties, express or implied, including any warranty of merchantability or fitness for a particular purpose. Use of RT/duroid ®microwave laminates in your particular application may yield different results.

These products may require an export license issued by the U.S. Department of Commerce for export of these materials from the United States or Canada.

RT/duroid and Duroid are registered trademark of Rogers Corporation for its microwave laminates.

Made under the following U.S. patents: 4,518,737; 4,335,180

ROGERS

Rogers Corporation
Microwave and Circuit Materials Division, 100 S. Roosevelt Avenue, Chandler, AZ 85226 Tel. 602 961-1382

Printed in U.S.A.

Issue 9/95

8334-095-5.0-HG

DATA
RT 1.6002

RT/duroid® 6002 PTFE Composite Circuit Board Material for Microwave Applications

RT/duroid® 6002 microwave material is the first low loss and low dielectric constant laminate to offer superior electrical and mechanical properties essential in designing complex microwave structures which are mechanically reliable and electrically stable.

The thermal coefficient of dielectric constant is extremely low from - 55°C to +150°C which provides the designers of filters, oscillators and delay lines the electrical stability in today's demanding applications.

A low Z axis coefficient of thermal expansion (CTE) ensures excellent reliability of plated through-holes. RT/duroid 6002 materials have been successfully temperature cycled (-55°C to 125°C) for over 5000 cycles without a single via failure.

Excellent dimensional stability (0.2 to 0.5 mils/inch) is achieved by matching the X and Y coefficient of expansion to copper. This often eliminates double etching to achieve tight positional tolerances.

The low tensile modulus (X,Y) greatly reduces the stress applied to solder joints and allows the expansion of the laminate to be constrained by a minimum amount of low CTE metal (6 ppm/°C) further increasing surface mount reliability.

1/4 oz to 2 oz./ft.^{2} electrodeposited copper, or 1/2 oz. to 2 oz/ft.^{2} rolled copper foil may be specified as cladding on dielectric thicknesses from 0.005" to 0.120". RT/duroid 6002 is also available clad with aluminum, brass, or copper plates.

Applications particularly suited to the unique properties of RT/duroid 6002 include flat and non-planar structures such as antennas, complex multilayer circuits with interlayer connections, and microwave circuits for aerospace designs in hostile environments. RT/duroid 6002 laminates have Underwriters Laboratories recognition under classification 94V-0 (Vertical Flammability Test).

(see reverse for product data)

ROGERS

RT 1.6002

RT/duroid® 6002 PTFE Composite

PROPERTY	TEST METHOD	CONDITIONS	UNITS(1)	DIRECTION	TYPICAL VALUE (2)
Dielectric Constant, ε_r	IPC-TM-650, 2.5.5.5	10 GHz/23°C	---	Z	2.94±0.04
Thermal Coefficient of ε_r	IPC-TM-650, 2.5.5.5	10 GHz/0-100°C	ppm/°C	Z	+16
Dissipation Factor, Tan δ	IPC-TM-650, 2.5.5.5	10 GHz/23°C	---	Z	0.0012
Volume Resistivity	ASTM D257	A	Mohm•cm	Z	10^6
Surface Resistivity	ASTM D257	A	Mohm	Z	10^7
Tensile Modulus	ASTM D638	23°C	MPa (kpsi)	X,Y	828 (120)
Ultimate Stress			MPa (kpsi)	X,Y	6.9 (1.0)
Ultimate Strain			%	X,Y	7.3
Compressive Modulus	Estimated		MPa (kpsi)	Z	2482 (360)
Water Absorption	IPC-TM-650, 2.6.2.1	D23/24	%	---	0.1
	ASTM D570	D48/50	%	---	0.13 max.
Specific Gravity (Ref. to Water)	ASTM D792	23°C	---	---	2.1
Specific Heat	Calculated	---	J/g/K (BTU/lb/°F)	---	0.93 (0.22)
Thermal Conductivity	ASTM C518	80°C	W/m/°K	---	0.60
Coefficient of Thermal Expansion	ASTMD3386	(10K/min)	ppm/°C	X,Y	16
				Z	24

(1) S1 units given first, with other frequently used units in parentheses.
(2) References: Internal TRs 3824,5016, 5017, 5035. Tests were at 23°C unless otherwise noted.

The above information is not intended to and does not create any warranties, express or implied including any warranty of merchantability or fitness for a particular purpose. Use of RT/duroid® microwave laminates in your particular application may yield different results.

ROGERS

Rogers Corporation
Microwave and Circuit Materials Division
100 S. Roosevelt Avenue, Chandler, AZ 85226-3415
602 961-1382 FAX: 602 961-4533
WEBSITE: http://www.rogers-corp.com/mwu/
ISO-9002 CERTIFIED

RT/duroid is a registered trademark of Rogers Corporation for its microwave laminates.
These products may require an export license issued by the United States Department of Commerce for export of these materials from the United States or Canada.

 Printed in U.S.A. Revised 2/97 0270.027.5.0-ON

TLC

APPLICATIONS

LNBs
PCS/PCN Large Format Antennas
Power Amplifiers
Passive Components

Taconic has over 35 years of experience coating fiberglass fabric with PTFE (polytetrafluoroethylene). This enables Taconic to manufacture copper clad PTFE/woven glass laminates with exceptionally well controlled electrical and mechanical properties.

Taconic TLC laminates are engineered to provide a cost effective substrate. TLC laminates are suitable for a wide range of microwave applications. TLC laminates offer far superior electrical performance compared to thermoset laminates (e.g. FR-4, PPO, BT, polyimide, and cyanate ester). The construction of the laminate provides exceptional mechanical stability. The dielectric constant (Dk) is typically offered at 3.2 +/-.05.

See "How to Order" on back page for a complete product listing.

TLC laminates can be sheared, drilled, milled, and plated using standard methods for PTFE/woven fiberglass materials. The laminates are dimensionally stable, and exhibit virtually no moisture absorption during fabrication processes.

TLC laminates are generally ordered clad on one or both sides with 1/2, 1, or 2 oz. electrodeposited copper foil. Contact our Customer Service Department for alternate claddings.

TLC laminates are tested in accordance with IPC-TM 650. A Certificate of Compliance containing actual test data accompanies each shipment.

TLC-32 TYPICAL VALUES					
Property	Test Method	Units	Value	Units	Value
Dielectric Constant @ 10 GHz	IPC-TM 650 2.5.5.5		3.20		3.20
Dissipation Factor @ 10 GHz	IPC-TM 650 2.5.5.5		0.0030		0.0030
Moisture Absorption	IPC-TM 650 2.6.2.1	%	<.02	%	<.02
Dielectric Breakdown	IPC-TM 650 2.5.6	kV	>60	kV	>60
Volume Resistivity	IPC-TM 650 2.5.17.1	Mohm/cm	10^7	Mohm/cm	10^7
Surface Resistivity	IPC-TM 650 2.5.17.1	Mohm	10^7	Mohm	10^7
Arc Resistance	IPC-TM 650 2.5.1	seconds	>180	seconds	>180
Flexural Strength Lengthwise	IPC-TM 650 2.4.4	lbs./in.	>40,000	N/mm²	>276
Flexural Strength Crosswise	IPC-TM 650 2.4.4	lbs./in.	>35,000	N/mm²	>241
Peel Strength (1oz copper)	IPC-TM 650 2.4.8	lbs./linear in.	12.0	N/mm	2.1
Thermal Conductivity	Cenco-Fitch	BTU/in./hr/ft²/°F	1.60	W/m/K	0.23
x-y CTE	ASTM D 3386 (TMA)	ppm/°C	9-12	ppm/°C	9-12
z CTE	ASTM D 3386 (TMA)	ppm/°C	70	ppm/°C	70
UL-94 Flammability Rating	UL-94		V-0		V-0

Type		Dk
TLY-5A		2.17
TLY-5		2.20
TLY-3		2.33
TLT-0	TLX-0	2.45
TLT-9	TLX-9	2.50
TLT-8	TLX-8	2.55
TLT-7	TLX-7	2.60
TLT-6	TLX-6	2.65
TLE-95		2.95
TLC-27		2.75
TLC-30		3.00
TLC-32		3.20
RF-35		3.50
CER-10		10

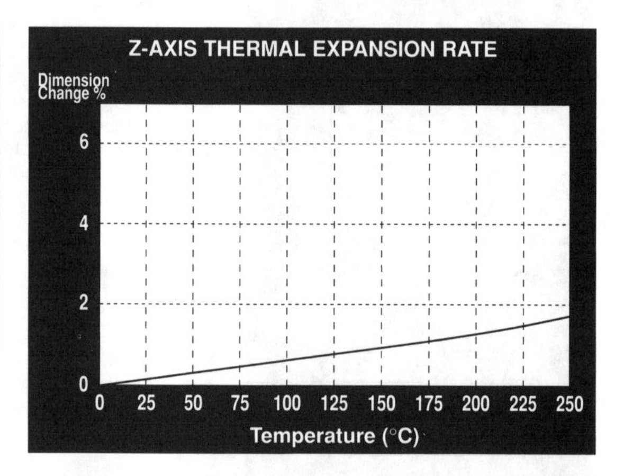

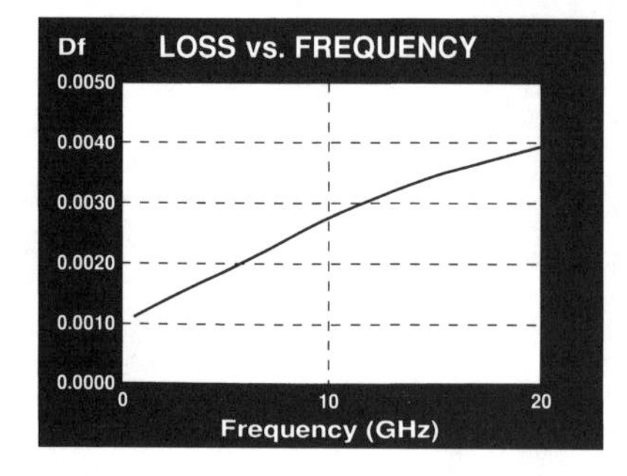

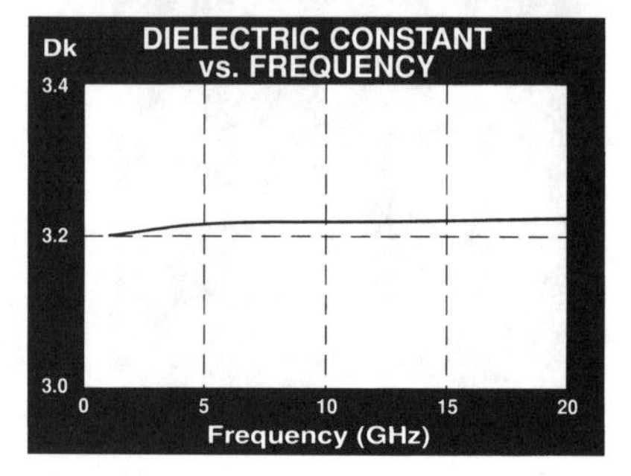

How to Order

Designation	Dielectric Constant	Dielectric Thickness	Dielectric Thickness
TLC - 27	2.75 +/– .05	.0145"	0.37mm
TLC - 30	3.00 +/– .05	.0200"	0.50mm
		.0300" - .0620"	0.78mm – 1.5mm
TLC - 32	3.20 +/– .05	≥ .0300"	≥ 0.78mm

Standard sheet size is 36" x 48" (914mm x 1220mm). Please contact our Customer Service Department for the availability of other sizes and claddings.

TLC can be ordered with the following electrodeposited copper:

Designation	Weight	Copper Thickness	Copper Thickness
CH	1/2 oz./sq. ft.	~ .0007"	~18 µm
C1	1 oz./sq. ft.	~ .0014"	~ 35 µm
C2	2 oz./sq. ft.	~ .0028"	~ 70 µm

Panels may be ordered cut to size

Typical Panel Sizes	
12" x 18"	304mm x 457mm
16" x 18"	406mm x 457mm
18" x 24"	457mm x 610mm
16" x 36"	406mm x 914mm
24" x 36"	610mm x 914mm
18" x 48"	457mm x 1220mm

An example of our part number is: TLC-32-0620-CH/CH-18" x 24" (TLC-32-0620-CH/CH-457mm x 610mm)

ADVANCED DIELECTRIC DIVISION

P.O. Box 69 • 136 Coonbrook Road
Petersburgh, New York 12138 • USA
TEL: 518-658-3202 • FAX: 518-658-3988
TOLL FREE: 800-833-1805 • FAX: 800-272-2503

Lynn Industrial Park
Mullingar, Co. Westmeath,
Republic of Ireland
TEL: +353-44-40477 • FAX: +353-44-44369

Na-906-2, APT Factory 148
Yatap-dong, Bundang-ku
Sungnam-si, Kyung'gi-do, Republic of Korea
TEL: +82-342-704-1858/9 • FAX: +82-342-704-1857

TLE

Microwave Radios
High Speed Digital Work Stations
Satellite Antenna Systems
Passive Components
High Layer Count MLBs
High Speed Chip Test MLBs

Taconic has over 35 years of experience coating fiberglass fabric with PTFE (polytetrafluoroethylene). This enables Taconic to manufacture copper clad PTFE/woven glass laminates with exceptionally well controlled electrical and mechanical properties.

Taconic TLE laminates are engineered to provide electrical, and mechanical properties to meet the requirements of complex microwave and high speed digital applications. The low Z-axis CTE of TLE laminates provides excellent plated through hole reliability. The low thermal expansion properties in the X and Y plane ensure high reliability in surface mount applications. The dielectric constant (Dk) exhibits minimal change over temperature. The Dk is typically offered at 2.95, with a tolerance of ±.05.

See "How to Order" on back page for a complete product listing.

TLE laminates can be sheared, drilled, milled, and plated using standard methods for PTFE/woven fiberglass materials. The laminates are dimensionally stable, and exhibit virtually no moisture absorption during fabrication processes.

TLE laminates are generally ordered clad on one or both sides with 1/2, 1, or 2 oz. electrodeposited copper foil. Contact our Customer Service Department for alternate claddings.

TLE laminates are tested in accordance with IPC-TM 650. A Certificate of Compliance containing actual test data accompanies each shipment.

TLE-95 TYPICAL VALUES					
Property	**Test Method**	**Units**	**Value**	**Units**	**Value**
Dielectric Constant @ 10 GHz	IPC-TM 650 2.5.5.5		2.95		2.95
Dissipation Factor @ 10 GHz	IPC-TM 650 2.5.5.5		0.0028		0.0028
Moisture Absorption	IPC-TM 650 2.6.2.1	%	<.02	%	<.02
Dielectric Breakdown	IPC-TM 650 2.5.6	kV	>60	kV	>60
Volume Resistivity	IPC-TM 650 2.5.17.1	Mohm/cm	10^7	Mohm/cm	10^7
Surface Resistivity	IPC-TM 650 2.5.17.1	Mohm	10^7	Mohm	10^7
Arc Resistance	IPC-TM 650 2.5.1	seconds	>180	seconds	>180
Flexural Strength Lengthwise	IPC-TM 650 2.4.4	lbs./in.	>35,000	N/mm^2	>241
Flexural Strength Crosswise	IPC-TM 650 2.4.4	lbs./in.	>30,000	N/mm^2	>207
Peel Strength (1oz copper)	IPC-TM 650 2.4.8	lbs./linear in.	12.0	N/mm	2.1
Thermal Conductivity	Cenco-Fitch	BTU/in./hr/ft^2/°F	1.60	W/m/K	0.23
x-y CTE	ASTM D 3386 (TMA)	ppm/°C	9-12	ppm/°C	9-12
z CTE	ASTM D 3386 (TMA)	ppm/°C	70	ppm/°C	70
UL-94 Flammability Rating	UL-94		V-0		V-0

Type		Dk
TLY-5A		2.17
TLY-5		2.20
TLY-3		2.33
TLT-0	TLX-0	2.45
TLT-9	TLX-9	2.50
TLT-8	TLX-8	2.55
TLT-7	TLX-7	2.60
TLT-6	TLX-6	2.65
TLE-95		2.95
TLC-27		2.75
TLC-30		3.00
TLC-32		3.20
RF-35		3.50
CER-10		10

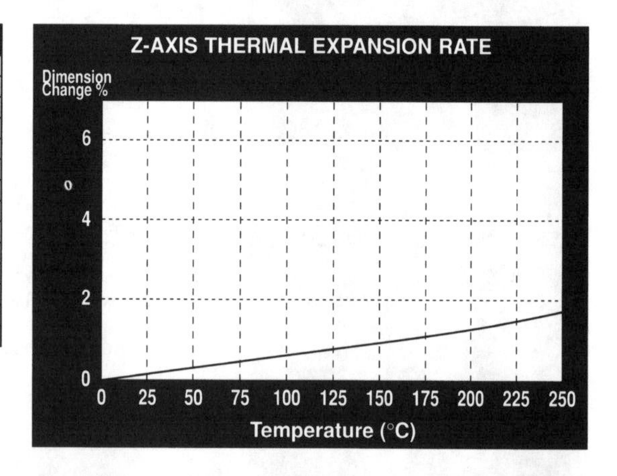

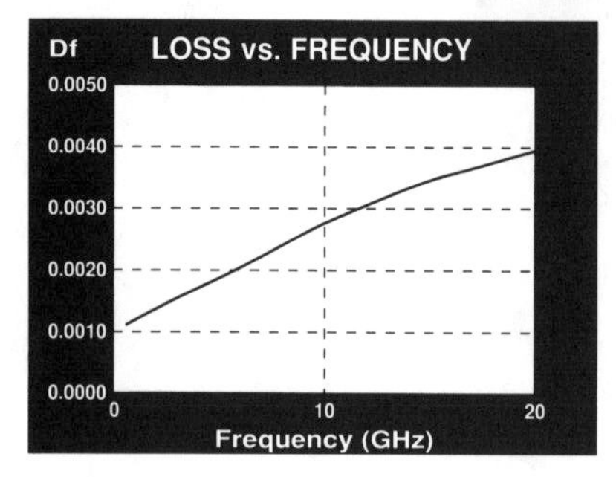

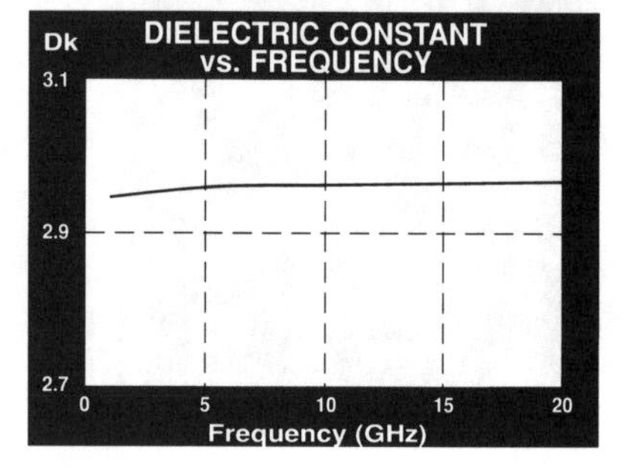

How to Order

Designation	Dielectric Constant	Dielectric Thickness	Dielectric Thickness
TLE - 95	3.00 +/– .05	.0052"	0.13mm
TLE - 95	2.95 +/– .05	.0100" - .0190"	0.25mm – 0.48mm
		.0200" - .0300"	0.50mm – 0.76mm
		≥ .0310"	≥ 0.80mm

Standard sheet size is 36" x 48" (914mm x 1200mm). Please contact our Customer Service Department for the availability of other sizes and claddings.

TLE can be ordered with the following electrodeposited copper:

Designation	Weight	Copper Thickness	Copper Thickness
CH	1/2 oz./sq. ft.	~ .0007"	~18 µm
C1	1 oz./sq. ft.	~ .0014"	~ 35 µm
C2	2 oz./sq. ft.	~ .0028"	~ 70 µm

Panels may be ordered cut to size

Typical Panel Sizes	
12" x 18"	304mm x 457mm
16" x 18"	406mm x 457mm
18" x 24"	457mm x 610mm
16" x 36"	406mm x 914mm
24" x 36"	610mm x 914mm
18" x 48"	457mm x 1220mm

An example of our part number is: TLE-95-0100-CH/CH-18" x 24" (TLE-95-0100-CH/CH-457mm x 610mm)

ADVANCED DIELECTRIC DIVISION

P.O. Box 69 • 136 Coonbrook Road
Petersburgh, New York 12138 • USA
TEL: 518-658-3202 • FAX: 518-658-3988
TOLL FREE: 800-833-1805 • FAX: 800-272-2503

Lynn Industrial Park
Mullingar, Co. Westmeath,
Republic of Ireland
TEL: +353-44-40477 • FAX: +353-44-44369

Na-906-2, APT Factory 148
Yatap-dong, Bundang-ku
Sungnam-si, Kyung'gi-do, Republic of Korea
TEL: +82-342-704-1858/9 • FAX: +82-342-704-1857

TLT

Taconic has over 35 years of experience coating fiberglass fabric with PTFE (polytetrafluoroethylene). This enables Taconic to manufacture copper clad PTFE/woven glass laminates with exceptionally well controlled electrical and mechanical properties.

These attributes effectively extend the usefulness of these laminates to X-Band and above.

The dielectric constant (Dk) range is 2.45 to 2.65. The Dk can be specified anywhere within this range with a tolerance of ±.05. The dissipation factor is approximately .0006 when measured @ 1 MHz.

See "How to Order" on back page for a complete product listing.

TLT laminates can be sheared, drilled, milled and plated using standard methods for PTFE/woven fiberglass materials. The laminates are dimensionally stable, and exhibit virtually no moisture absorption during fabrication processes.

TLT laminates are generally ordered clad on one or both sides with 1/2, 1, or 2 oz. electrodeposited copper. Contact our Customer Service Department for alternate claddings.

Typical applications for TLT laminates include radar systems, phased array antennas, mobile communication systems, microwave test equipment, microwave transmission devices and RF components.

TLT laminates are tested in accordance with IPC-TM 650. A Certificate of Compliance containing actual test data accompanies each shipment.

TLT-9 TYPICAL VALUES					
Property	Test Method	Units	Value	Units	Value
Dielectric Constant @ 1 MHz	IPC-TM 650 2.5.5.3		2.50		2.50
Dissipation Factor @ 1 MHz	IPC-TM 650 2.5.5.3		0.0006		0.0006
Moisture Absorption	IPC-TM 650 2.6.2.1	%	<.02	%	<.02
Dielectric Breakdown	IPC-TM 650 2.5.6	kV	>60	kV	>60
Volume Resistivity	IPC-TM 650 2.5.17.1	Mohm/cm	10^7	Mohm/cm	10^7
Surface Resistivity	IPC-TM 650 2.5.17.1	Mohm	10^7	Mohm	10^7
Arc Resistance	IPC-TM 650 2.5.1	seconds	>180	seconds	>180
Flexural Strength Lengthwise	IPC-TM 650 2.4.4	lbs./in.	>23,000	N/mm²	>159
Flexural Strength Crosswise	IPC-TM 650 2.4.4	lbs./in.	>19,000	N/mm²	>131
Peel Strength (1oz copper)	IPC-TM 650 2.4.8	lbs./linear in.	12.0	N/mm	2.1
Thermal Conductivity	Cenco-Finch	BTU/in./hr/ft²/°F	2.34	W/m/K	0.34
x-y CTE	ASTM D 3385 (TMA)	ppm/°C	9-12	ppm/°C	9-12
z CTE	ASTM D 3385 (TMA)	ppm/°C	130-145	ppm/°C	130-145
UL-94 Flammability Rating	UL-94		V-0		V-0

Type		Dk
TLY-5A		2.17
TLY-5		2.20
TLY-3		2.33
TLT-0	TLX-0	2.45
TLT-9	TLX-9	2.50
TLT-8	TLX-8	2.55
TLT-7	TLX-7	2.60
TLT-6	TLX-6	2.65
TLE-95		2.95
TLC-27		2.75
TLC-30		3.00
TLC-32		3.20
RF-35		3.50
CER-10		10

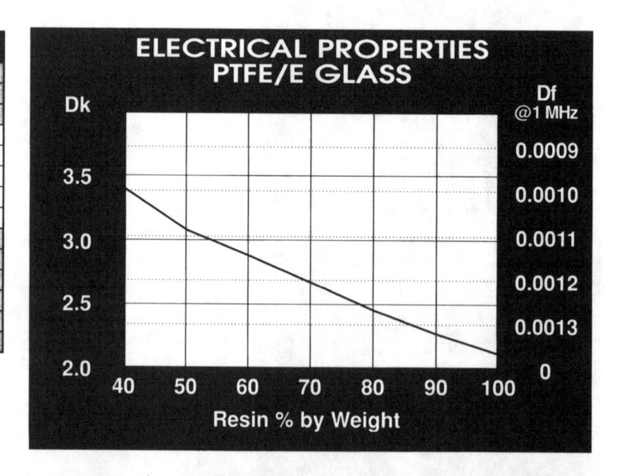

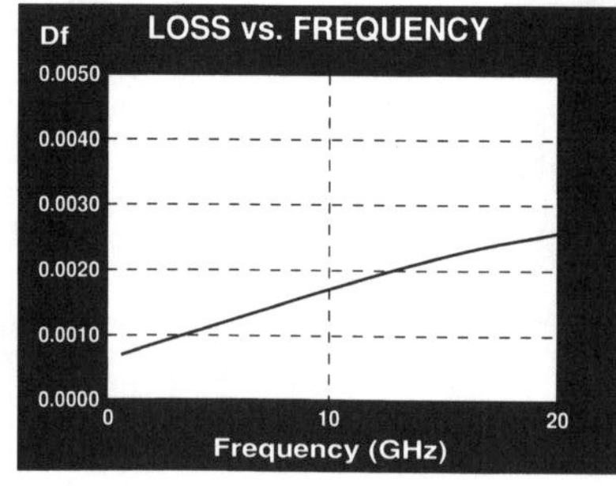

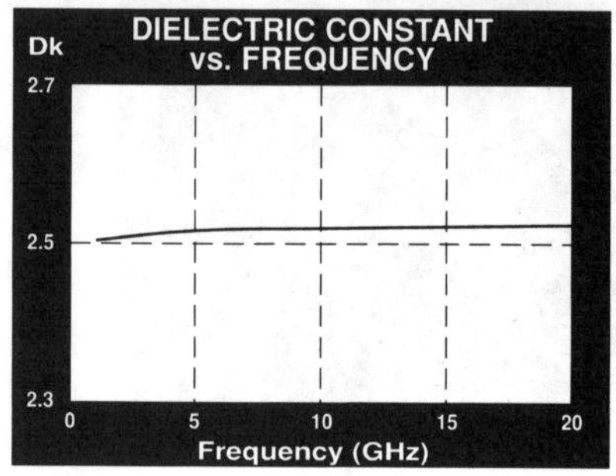

How to Order

Designation	Dielectric Constant	Dielectric Thickness	Dielectric Thickness
TLT - 0	2.45 +/– .05	.0050" - .0190"	0.13mm – 0.48mm
		.0200" - .0300"	0.50mm – 0.76mm
		≥ .0310"	≥ 0.80mm
TLT - 9	2.50 +/– .05	.0050" - .0190"	0.13mm – 0.48mm
		.0200" - .0300"	0.50mm – 0.76mm
		≥ .0310"	≥ 0.80mm
TLT - 8	2.55 +/– .05	.0050" - .0190"	0.13mm – 0.48mm
		.0200" - .0300"	0.50mm – 0.76mm
		≥ .0310"	≥ 0.80mm
TLT - 7	2.60 +/– .05	.0050" - .0190"	0.13mm – 0.48mm
		.0200" - .0300"	0.50mm – 0.76mm
		≥ .0310"	≥ 0.80mm
TLT - 6	2.65 +/– .05	.0050" - .0190"	0.13mm – 0.48mm
		.0200" - .0300"	0.50mm – 0.76mm
		≥ .0310"	≥ 0.80mm

Standard sheet size is 36" x 48" (914mm x 1220mm). Please contact our Customer Service Department for the availability of other sizes and claddings.

TLT can be ordered with the following electrodeposited copper:

Designation	Weight	Copper Thickness	Copper Thickness
CH	1/2 oz./sq. ft.	~ .0007"	~18 µm
C1	1 oz./sq. ft.	~ .0014"	~ 35 µm
C2	2 oz./sq. ft.	~ .0028"	~ 70 µm

Panels may be ordered cut to size

Typical Panel Sizes	
12" x 18"	304mm x 457mm
16" x 18"	406mm x 457mm
18" x 24"	457mm x 610mm
16" x 36"	406mm x 914mm
24" x 36"	610mm x 914mm
18" x 48"	457mm x 1220mm

An example of our part number is: TLT-9-0310-CH/CH-18" x 24" (TLT-9-0310-CH/CH-457mm x 610mm)

ADVANCED DIELECTRIC DIVISION

P.O. Box 69 • 136 Coonbrook Road
Petersburgh, New York 12138 • USA
TEL: 518-658-3202 • FAX: 518-658-3988
TOLL FREE: 800-833-1805 • FAX: 800-272-2503

Lynn Industrial Park
Mullingar, Co. Westmeath,
Republic of Ireland
TEL: +353-44-40477 • FAX: +353-44-44369

Na-906-2, APT Factory 148
Yatap-dong, Bundang-ku
Sungnam-si, Kyung'gi-do, Republic of Korea
TEL: +82-342-704-1858/9 • FAX: +82-342-704-1857

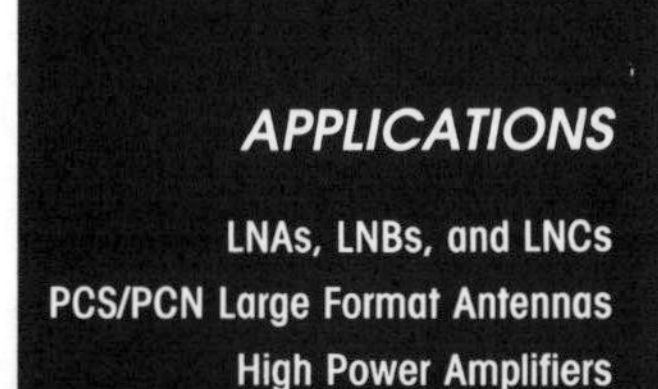

TLX

Taconic has over 35 years of experience coating fiberglass fabric with PTFE (polytetrafluoroethylene). This enables Taconic to manufacture copper clad PTFE/woven glass laminates with exceptionally well controlled electrical and mechanical properties.

The dielectric constant (Dk) range is 2.45 to 2.65. The Dk can be specified anywhere within this range with a tolerance of ±.04. The dissipation factor is approximately .0019 when measured at 10 GHz.

See "How to Order" on back page for a complete product listing.

TLX laminates can be sheared, drilled, milled and plated using standard methods for PTFE/woven fiberglass materials. The laminates are dimensionally stable, and exhibit virtually no moisture absorption during fabrication processes.

TLX laminates are generally ordered clad on one or both sides with 1/2, 1, or 2 oz. electrodeposited copper. Contact our Customer Service Department for alternate claddings.

Typical applications for TLX laminates include radar systems, phased array antennas, mobile communication systems, microwave test equipment, microwave transmission devices and RF components.

TLX laminates are tested in accordance with IPC-TM 650. A Certificate of Compliance containing actual test data accompanies each shipment.

TLX-9 TYPICAL VALUES					
Property	**Test Method**	**Units**	**Value**	**Units**	**Value**
Dielectric Constant @ 10 GHz	IPC-TM 650 2.5.5.5		2.50		2.50
Dissipation Factor @ 10 GHz	IPC-TM 650 2.5.5.5		0.0019		0.0019
Moisture Absorption	IPC-TM 650 2.6.2.1	%	<.02	%	<.02
Dielectric Breakdown	IPC-TM 650 2.5.6	kV	>60	kV	>60
Volume Resistivity	IPC-TM 650 2.5.17.1	Mohm/cm	10^7	Mohm/cm	10^7
Surface Resistivity	IPC-TM 650 2.5.17.1	Mohm	10^7	Mohm	10^7
Arc Resistance	IPC-TM 650 2.5.1	seconds	>180	seconds	>180
Flexural Strength Lengthwise	IPC-TM 650 2.4.4	lbs./in.	>23,000	N/mm²	>159
Flexural Strength Crosswise	IPC-TM 650 2.4.4	lbs./in.	>19,000	N/mm²	>131
Peel Strength (1oz copper)	IPC-TM 650 2.4.8	lbs./linear in.	12.0	N/mm	2.1
Thermal Conductivity	Cenco-Fitch	BTU/in./hr/ft²/°F	2.34	W/m/K	0.34
x-y CTE	ASTM D 3386 (TMA)	ppm/°C	9-12	ppm/°C	9-12
z CTE	ASTM D 3386 (TMA)	ppm/°C	130-145	ppm/°C	130-145
UL-94 Flammability Rating	UL-94		V-0		V-0

Type		Dk
TLY-5A		2.17
TLY-5		2.20
TLY-3		2.33
TLT-0	TLX-0	2.45
TLT-9	TLX-9	2.50
TLT-8	TLX-8	2.55
TLT-7	TLX-7	2.60
TLT-6	TLX-6	2.65
TLE-95		2.95
TLC-27		2.75
TLC-30		3.00
TLC-32		3.20
RF-35		3.50
CER-10		10

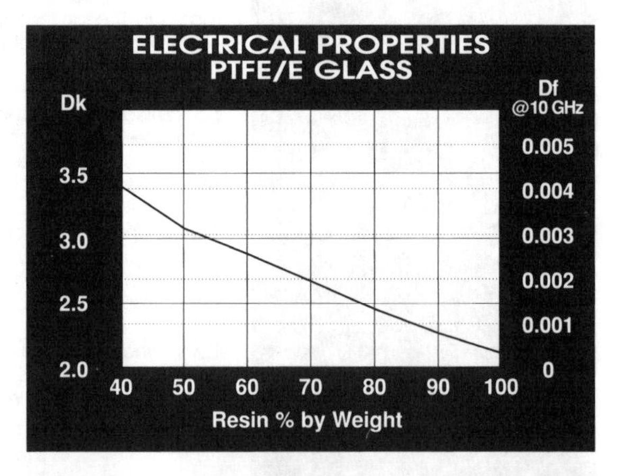

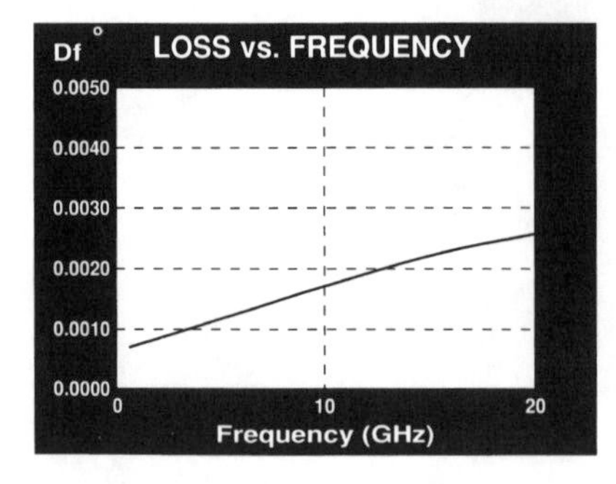

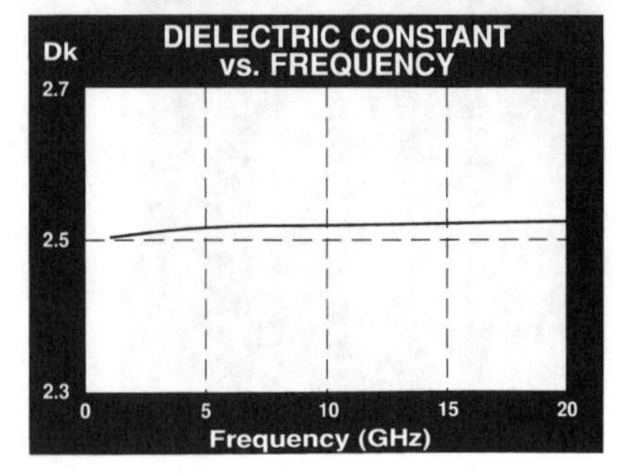

How to Order

Designation	Dielectric Constant	Dielectric Thickness	Dielectric Thickness
TLX - 0	2.45 +/– .04	.0050" - .0190"	0.13mm – 0.48mm
		.0200" - .0300"	0.50mm – 0.76mm
		≥ .0310"	≥ 0.80mm
TLX - 9	2.50 +/– .04	.0050" - .0190"	0.13mm – 0.48mm
		.0200" - .0300"	0.50mm – 0.76mm
		≥ .0310"	≥ 0.80mm
TLX - 8	2.55 +/– .04	.0050" - .0190"	0.13mm – 0.48mm
		.0200" - .0300"	0.50mm – 0.76mm
		≥ .0310"	≥ 0.80mm
TLX - 7	2.60 +/– .04	.0050" - .0190"	0.13mm – 0.48mm
		.0200" - .0300"	0.50mm – 0.76mm
		≥ .0310"	≥ 0.80mm
TLX - 6	2.65 +/– .04	.0050" - .0190"	0.13mm – 0.48mm
		.0200" - .0300"	0.50mm – 0.76mm
		≥ .0310"	≥ 0.80mm

Standard sheet size is 36" x 48" (914mm x 1220mm). Please contact our Customer Service Department for the availability of other sizes and claddings.

TLX can be ordered with the following electrodeposited copper:

Designation	Weight	Copper Thickness	Copper Thickness
CH	1/2 oz./sq. ft.	~ .0007"	~18 μm
C1	1 oz./sq. ft.	~ .0014"	~ 35 μm
C2	2 oz./sq. ft.	~ .0028"	~ 70 μm

Panels may be ordered cut to size

Typical Panel Sizes	
12" x 18"	304mm x 457mm
16" x 18"	406mm x 457mm
18" x 24"	457mm x 610mm
16" x 36"	406mm x 914mm
24" x 36"	610mm x 914mm
18" x 48"	457mm x 1220mm

An example of our part number is: TLX-9-0310-CH/CH-18" x 24" (TLX-9-0310-CH/CH-457mm x 610mm)

ADVANCED DIELECTRIC DIVISION

P.O. Box 69 • 136 Coonbrook Road
Petersburgh, New York 12138 • USA
TEL: 518-658-3202 • FAX: 518-658-3988
TOLL FREE: 800-833-1805 • FAX: 800-272-2503

Lynn Industrial Park
Mullingar, Co. Westmeath,
Republic of Ireland
TEL: +353-44-40477 • FAX: +353-44-44369

Na-906-2, APT Factory 148
Yatap-dong, Bundang-ku
Sungnam-si, Kyung'gi-do, Republic of Korea
TEL: +82-342-704-1858/9 • FAX: +82-342-704-1857

APPLICATIONS

Automotive Radars
Satellite Communications
Cellular Communications
Power Amplifiers
LNBs, LNAs, LNCs
Aerospace:
Radars
Guidance Telemetry
High Frequency Communication
Military

TLY

Taconic has over 35 years of experience coating fiberglass fabric with PTFE (polytetrafluoroethylene). This enables Taconic to manufacture copper clad PTFE/woven glass laminates with exceptionally well controlled electrical and mechanical properties.

Taconic TLY laminates are manufactured from woven fiberglass fabric coated wlth PTFE interleaved with thin sheets of pure PTFE. The woven matrix produces a more mechanically stable laminate with a more uniform dielectric constant (Dk) than traditional non-woven products. The exceptionally low dissipation factor extends the usefulness of this product to 35 GHz and above.

The dielectric constant range is 2.17 to 2.40. For most thicknesses, the dielectric constant can be specified anywhere within this range with a tolerance of ±.02. In the low dielectric constant range, the dissipation factor is approximately 0.0009 when measured at 10 GHz.

See "How to Order" on back page for a complete product listing.

TLY laminates can be sheared, drilled, milled and plated using the accepted methods for PTFE/woven fiberglass laminates. The laminates are dimensionally stable and are resistant to the solvents and reagents used during fabrication. TLY laminates are being used in composite multi-layer circuitboard applications.

Typical applications for TLY laminates include mobile communications systems, microwave transmission devices, phased array antennas, RF components, and automotive and military radars.

TLY laminates are generally ordered clad on one or both sides with 1/2, 1, and 2 oz. electrodeposited copper. Contact our Customer Service Department for alternate claddings.

TLY laminates are tested in accordance with IPC-TM 650. A certificate of compliance containing lot-specific test data accompanies each shipment.

TLY-5A TYPICAL VALUES					
Property	Test Method	Units	Value	Units	Value
Dielectric Constant @ 10 GHz	IPC-TM 650 2.5.5.5		2.17		2.17
Dissipation Factor @ 10 GHz	IPC-TM 650 2.5.5.5		0.0009		0.0009
Moisture Absorption	IPC-TM 650 2.6.2.1	%	<.02	%	<.02
Dielectric Breakdown	IPC-TM 650 2.5.6	kV	>60	kV	>60
Volume Resistivity	IPC-TM 650 2.5.17.1	Mohm/cm	10^7	Mohm/cm	10^7
Surface Resistivity	IPC-TM 650 2.5.17.1	Mohm	10^7	Mohm	10^7
Arc Resistance	IPC-TM 650 2.5.1	seconds	>180	seconds	>180
Flexural Strength Lengthwise	IPC-TM 650 2.4.4	lbs./in.	>12,000	N/mm²	>83
Flexural Strength Crosswise	IPC-TM 650 2.4.4	lbs./in.	>10,000	N/mm²	>69
Peel Strength (1oz copper)	IPC-TM 650 2.4.8	lbs./linear in.	12.0	N/mm	2.1
Thermal Conductivity	Cenco-Fitch	BTU/in./hr/ft²/°F	2.79	W/m/K	0.40
x-y CTE	ASTM D 3386 (TMA)	ppm/°C	20	ppm/°C	20
z CTE	ASTM D 3386 (TMA)	ppm/°C	280	ppm/°C	280
UL-94 Flammability Rating	UL-94		V-0		V-0

Type		Dk
TLY-5A		2.17
TLY-5		2.20
TLY-3		2.33
TLT-0	TLX-0	2.45
TLT-9	TLX-9	2.50
TLT-8	TLX-8	2.55
TLT-7	TLX-7	2.60
TLT-6	TLX-6	2.65
TLE-95		2.95
TLC-27		2.75
TLC-30		3.00
TLC-32		3.20
RF-35		3.50
CER-10		10

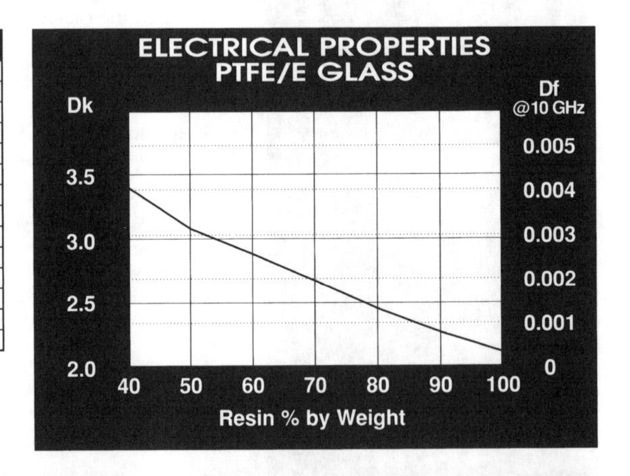

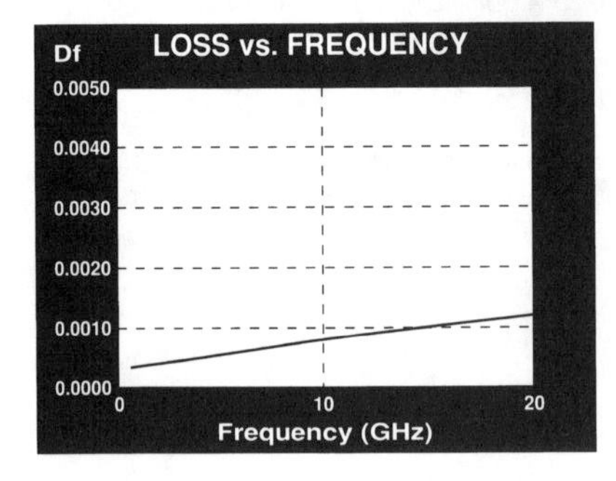

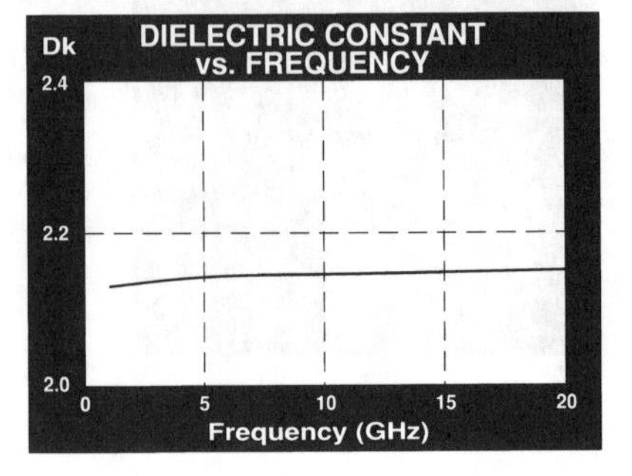

How to Order

Designation	Dielectric Constant	Dielectric Thickness	Dielectric Thickness
TLY - 5A	2.17 - +/– .02	≥ .0310"	≥ 0.80mm
TLY - 5	2.20 +/– .02	≥ .0050"	≥ 0.13mm
TLY - 3	2.33 +/– .02	≥ .0050"	≥ 0.13mm

Standard sheet size is 36" x 48" (914mm x 1220mm). Please contact our Customer Service Department for the availability of other sizes and claddings.

TLY can be ordered with the following electrodeposited copper:

Designation	Weight	Copper Thickness	Copper Thickness
CH	1/2 oz./sq. ft.	~ .0007"	~18 µm
C1	1 oz./sq. ft.	~ .0014"	~ 35 µm
C2	2 oz./sq. ft.	~ .0028"	~ 70 µm

Panels may be ordered cut to size

Typical Panel Sizes	
12" x 18"	304mm x 457mm
16" x 18"	406mm x 457mm
18" x 24"	457mm x 610mm
16" x 36"	406mm x 914mm
24" x 36"	610mm x 914mm
18" x 48"	457mm x 1220mm

An example of our part number is: TLY-5-0100-CH/CH-18" x 24" (TLY-5-0100-CH/CH-457mm x 610mm)

P.O. Box 69 • 136 Coonbrook Road
Petersburgh, New York 12138 • USA
TEL: 518-658-3202 • FAX: 518-658-3988
TOLL FREE: 800-833-1805 • FAX: 800-272-2503

Lynn Industrial Park
Mullingar, Co. Westmeath,
Republic of Ireland
TEL: +353-44-40477 • FAX: +353-44-44369

Na-906-2, APT Factory 148
Yatap-dong, Bundang-ku
Sungnam-si, Kyung'gi-do, Republic of Korea
TEL: +82-342-704-1858/9 • FAX: +82-342-704-1857

RF-35

APPLICATIONS

Power Amplifiers
Filters and Couplers
Passive Components

RF-35 is the best choice for low cost, high volume commercial microwave and radio frequency applications.

RF-35 has excellent peel strength for 1/2 ounce and 1 ounce copper (even in comparison to standard epoxy materials), a critical aspect whenever rework is required.

RF-35's Tg is over 600°F (315°C)

RF-35's ultra low moisture absorption rate and low dissipation factor minimize phase shift with frequency.

RF-35 is dimensionally stable due to the use of woven fabrics in its design.

RF-35 laminates are generally ordered clad on one or both sides with 1/2, 1, and 2 oz. electrodeposited copper.

RF-35 laminates exhibit flammability of V-0, and are tested in accordance with IPC-TM 650. A certificate of compliance containing lot-specific test data accompanies each shipment.

See "How to Order" on back page for a complete product listing.

RF-35 TYPICAL VALUES					
Property	**Test Method**	**Units**	**Value**	**Units**	**Value**
Dielectric Constant @ 1.9 GHz	IPC-TM 650 2.5.5		3.50		3.50
Dissipation Factor @ 1.9 GHz	IPC-TM 650 2.5.5		0.0018		0.0018
Moisture Absorption (.060")	IPC-TM 650 2.6.2.1	%	.02	%	.02
Peel Strength (1/2 oz. copper)	IPC-TM 650 2.4.8	lbs./linear inch	>8.0	N/mm	>1.5
Peel Strength (1 oz. copper)	IPC-TM 650 2.4.8	lbs./linear inch	>10.0	N/mm	>1.8
Dielectric Breakdown	IPC-TM 650 2.5.6	kV	41	kV	41
Volume Resistivity	IPC-TM 650 2.5.17.1	Mohm/cm	1.26×10^9	Mohm/cm	1.26×10^9
Surface Resistivity	IPC-TM 650 2.5.17.1	Mohm	1.46×10^8	Mohm	1.46×10^8
Arc Resistance	IPC TM 650 2.5.1	seconds	>180	seconds	>180
Flexural Strength Lengthwise	ASTM D 790	psi	>22,000	N/mm^2	>152
Flexural Strength Crosswise	ASTM D 790	psi	>18,000	N/mm^2	>124
Thermal Conductivity	Cenco-Fitch	BTU/in/hr/ft^2/°F	1.416	W/m/K	0.20
Tensile Strength Lengthwise	ASTM D 638	psi	27,000	N/mm^2	187
Tensile Strength Crosswise	ASTM D 638	psi	21,000	N/mm^2	145
Dimensional Stability Lengthwise	IPC-TM 650 2.4.39	in/in	.00004	mm/mm	.00004
Dimensional Stability Crosswise	IPC-TM 650 2.4.39	in/in	-.00010	mm/mm	-.00010
x-y CTE	ASTM D 3386 (TMA)	ppm/°C	19-24	ppm/°C	19-24
z CTE	ASTM D 3386 (TMA)	ppm/°C	64	ppm/°C	64
Flammability	UL-94		V-0		V-0
Hardness	Rockwell M Scale		34		34

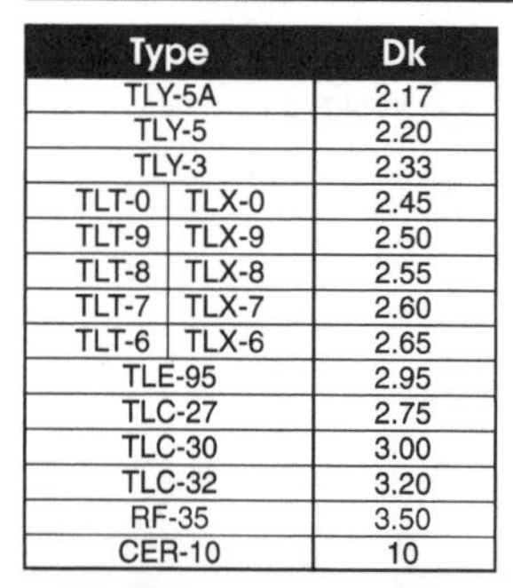

Type		Dk
TLY-5A		2.17
TLY-5		2.20
TLY-3		2.33
TLT-0	TLX-0	2.45
TLT-9	TLX-9	2.50
TLT-8	TLX-8	2.55
TLT-7	TLX-7	2.60
TLT-6	TLX-6	2.65
TLE-95		2.95
TLC-27		2.75
TLC-30		3.00
TLC-32		3.20
RF-35		3.50
CER-10		10

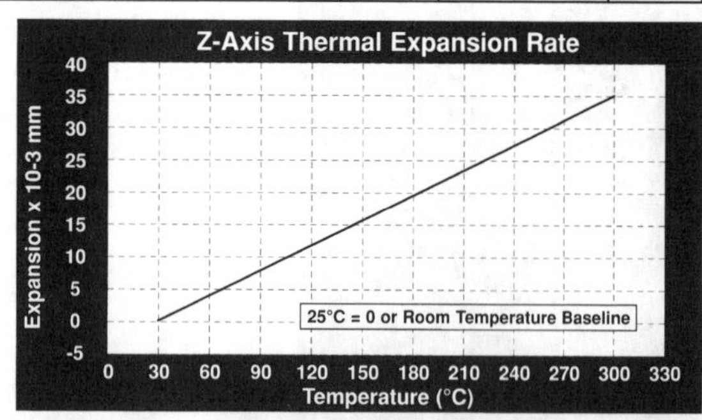

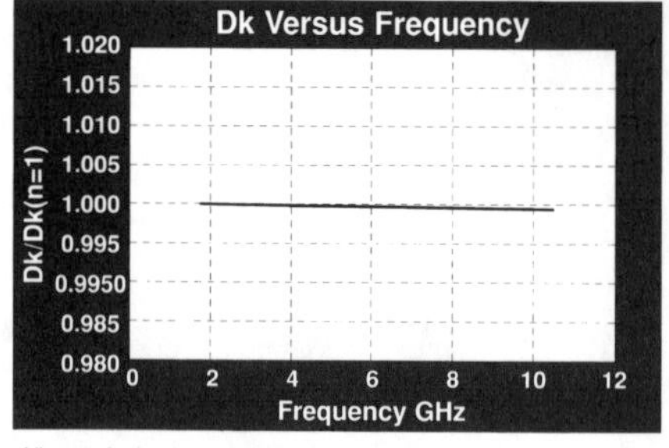

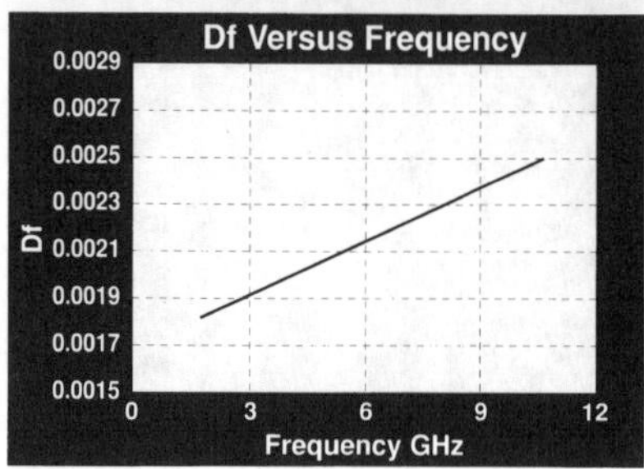

All reported values are typical and should not be used for specification purposes. In all instances the user shall determine suitability in any given application.

How to Order

Designation	Dielectric Constant
RF-35	3.5

Available Thickness	Available Thickness
0.0200"	0.50mm
0.0300"	0.76mm
0.0600"	1.52mm

Standard sheet size is 36" x 48" (914mm x 1220mm). Please contact our Customer Service Department for availability of other sizes and claddings.

RF-35 can be ordered with the following electrodeposited copper:

Designation	Weight	Copper Thickness	Copper Thickness
CH	1/2 oz./sq. ft.	~ .0007"	~18 µm
C1	1 oz./sq. ft.	~ .0014"	~ 35 µm
C2	2 oz./sq. ft.	~ .0028"	~ 70 µm

Panels may be ordered cut to size

Typical Panel Sizes	
12" x 18"	304mm x 457mm
16" x 18"	406mm x 457mm
18" x 24"	457mm x 610mm
16" x 36"	406mm x 914mm
24" x 36"	610mm x 914mm
18" x 48"	457mm x 1220mm

An example of our part number is: RF-35-0600-CH/CH-18" x 24" (RF-35-0600-CH/CH-457mm x 610mm)

ADVANCED DIELECTRIC DIVISION

P.O. Box 69 • 136 Coonbrook Road
Petersburgh, New York 12138 • USA
TEL: 518-658-3202 • FAX: 518-658-3988
TOLL FREE: 800-833-1805 • FAX: 800-272-2503

Lynn Industrial Park
Mullingar, Co. Westmeath,
Republic of Ireland
TEL: +353-44-40477 • FAX: +353-44-44369

Na-906-2, APT Factory 148
Yatap-dong, Bundang-ku
Sungnam-si, Kyung'gi-do, Republic of Korea
TEL: +82-342-704-1858/9 • FAX: +82-342-704-1857

CER-10

APPLICATIONS

Power Amplifiers

Filters and Couplers

Passive Components

CER-10 is a ceramic loaded Dk-10 laminate, based on a woven glass reinforcement. CER-10 is a result of Taconic's expertise in both ceramic fill technology and in coated PTFE fabrics.

CER-10 exhibits exceptional interlaminar bond strength and solder resistance. CER-10's proprietary composition results in low moisture absorption and uniform electrical properties.

CER-10's woven glass reinforcement ensures excellent dimensional stability and enhances flexural strength. This Dk-10 laminate exhibits low Z-axis expansion (CTE 46 ppm/°C), allowing for plated-through-hole reliability in extreme thermal environments.

CER-10 laminates can be sheared, drilled, milled and plated, using standard methods for PTFE/woven fiber-glass materials. For processing information, refer to the CER-10 Fabrication Guidelines.

CER-10 laminates are ordered clad on both sides with 1/2, 1, or 2 oz. electrodeposited copper foil. This laminate is available in standard master sheet sizes of 36" x 48". Various panel sizes are available. Contact our Customer Service Department for more information.

CER-10 laminates are tested in accordance with IPC-TM-650. A Certificate of Compliance containing actual test data accompanies each shipment.

See "How to Order" on back page for a complete product listing.

CER-10 TYPICAL VALUES

Property	Test Method	Units	Value	Units	Value
Dielectric Constant (Nominal)	IPC TM 650 2.5.5.6		10		10
Dissipation Factor @ X-band	IPC TM 650 2.5.5.5		0.0035		0.0035
Moisture Absorption	IPC TM 650 2.6.2.1	%	.02	%	.02
Dielectric Breakdown	IPC TM 650 2.5.6	kV	44	kV	44
Volume Resistivity	IPC TM 650 2.5.17.1	Mohm/cm	2.1×10^8	Mohm/cm	2.1×10^8
Surface Resistivity	IPC TM 650 2.5.17.1	Mohm	1.1×10^9	Mohm	1.1×10^9
Arc Resistance	IPC TM 650 2.5.1	Seconds	>180	Seconds	>180
Flexural Strength Lengthwise	ASTM D 790	psi	16,500	N/mm^2	114
Flexural Strength Crosswise	ASTM D 790	psi	15,500	N/mm^2	107
Peel Strength (1/2 oz. ed.)	IPC TM 650 2.4.8	lbs./linear inch	5.0	N/mm	0.88
Thermal Conductivity	ASTM C 177 (guarded hotplate)	W/m/°K	0.29	W/m/°K	0.29
		$BTU/in/hr/ft^2/°F$	0.168	$BTU/in/hr/ft^2/°F$	0.168
Tensile Strength Lengthwise	ASTM D 638	psi	7700	N/mm^2	532
Tensile Strength Crosswise	ASTM D 638	psi	6700	N/mm^2	46
Dimensional Stability Lengthwise	IPC TM 650 2.4.39	in/in	-.0002	mm/mm	-.002
Dimensional Stability Crosswise	IPC TM 650 2.4.39	in/in	-.0003	mm/mm	-.0003
x-y CTE	ASTM D 3386 (TMA)	ppm/°C	13-15	ppm/°C	13-15
z CTE	ASTM D 3386 (TMA)	ppm/°C	46	ppm/°C	46
Flammability Rating	UL-94		V-0		V-0
Specific Gravity		g/cm^3	3.05	g/cm^3	3.05

Type		Dk
TLY-5A		2.17
TLY-5		2.20
TLY-3		2.33
TLT-0	TLX-0	2.45
TLT-9	TLX-9	2.50
TLT-8	TLX-8	2.55
TLT-7	TLX-7	2.60
TLT-6	TLX-6	2.65
TLE-95		2.95
TLC-27		2.75
TLC-30		3.00
TLC-32		3.20
RF-35		3.50
CER-10		10

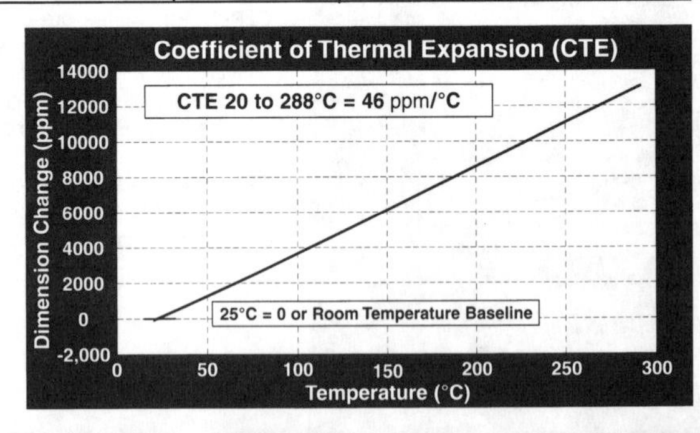

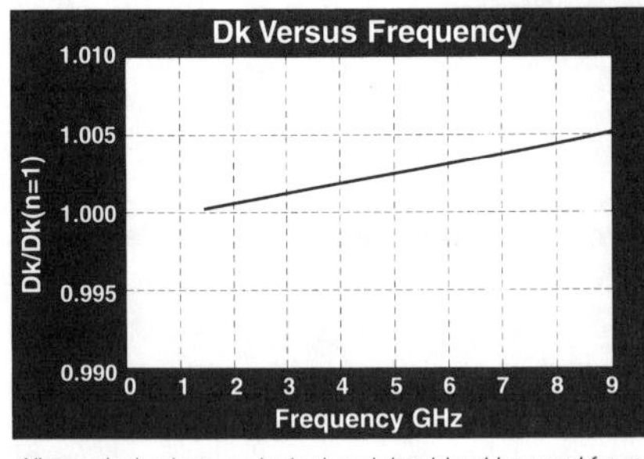

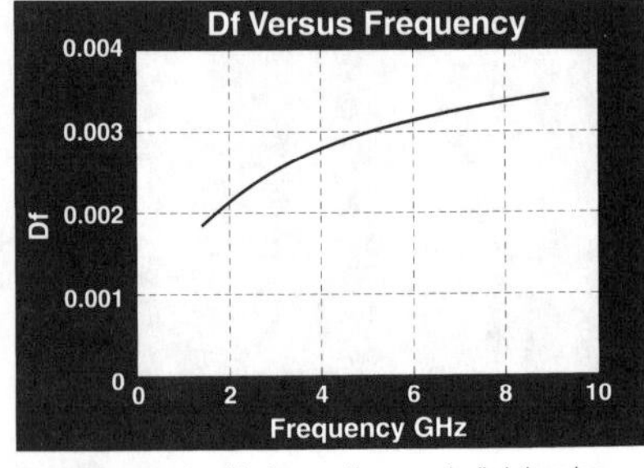

All reported values are typical and should not be used for specification purposes. In all instances the user shall determine suitability in any given application.

How to Order

Designation	Dielectric Constant
CER-10	10

Available Thickness	Available Thickness (mm)
0.0250"	0.63
0.0470"	1.19
0.0500"	1.27
0.0620"	1.57

Standard sheet size is 36" x 48" (914mm x 1220mm). Please contact our Customer Service Department for more information.

TLC can be ordered with the following electrodeposited copper:

Designation	Weight	Copper Thickness	Copper Thickness
CH	1/2 oz./sq. ft.	~ .0007"	~18 μm
C1	1 oz./sq. ft.	~ .0014"	~ 35 μm
C2	2 oz./sq. ft.	~ .0028"	~ 70 μm

Panels may be ordered cut to size

Typical Panel Sizes	
12" x 18"	304mm x 457mm
16" x 18"	406mm x 457mm
18" x 24"	457mm x 610mm
16" x 36"	406mm x 914mm
24" x 36"	610mm x 914mm
18" x 48"	457mm x 1220mm

An example of our part number is: CER-10-0620-CH/CH-18" x 24" (CER-10-0620-CH/CH-457mm x 610mm)

ADVANCED DIELECTRIC DIVISION

P.O. Box 69 • 136 Coonbrook Road
Petersburgh, New York 12138 • USA
TEL: 518-658-3202 • FAX: 518-658-3988
TOLL FREE: 800-833-1805 • FAX: 800-272-2503

Lynn Industrial Park
Mullingar, Co. Westmeath,
Republic of Ireland
TEL: +353-44-40477 • FAX: +353-44-44369

Na-906-2, APT Factory 148
Yatap-dong, Bundang-ku
Sungnam-si, Kyung'gi-do, Republic of Korea
TEL: +82-342-704-1858/9 • FAX: +82-342-704-1857

Temperature Stable Microwave Materials

TMM® Temperature Stable Microwave Composite Circuit Board Materials

TMM® Temperature Stable Microwave laminates are ceramic thermoset polymer composites designed for high reliability stripline and microstrip applications. TMM laminates are available in a wide range of dielectric constants and claddings.

The electrical and mechanical properties of TMM laminates combine many of the benefits of both ceramic and traditional PTFE microwave circuit laminates, without requiring the specialized production techniques common to these materials. TMM laminates do not require a sodium napthanate treatment prior to electroless plating.

An exceptionally low thermal coefficient of dielectric constant, typically less than 30 ppm per degree C, results in consistent electrical performance over broad temperature ranges.

TMM laminates have isotropic coefficients of thermal expansion, very closely matched to copper, allowing for production of high reliability plated through holes, and low etch shrinkage values. Furthermore, the thermal conductivity of TMM laminates is approximately twice that of traditional PTFE/ceramic laminates, facilitating heat removal.

TMM laminates are based on thermoset resins, and do not soften when heated. As a result, wire bonding of component leads to circuit traces can be performed without concerns of pad lifting or substrate deformation.

TMM laminates combine many of the desirable features of ceramic substrates with the ease of soft substrate processing techniques. TMM temperature stable microwave laminates are available clad with 1/4 oz/ft^2 to 2 oz/ ft^2 electrodeposited copper foil, or bonded directly to brass or aluminum plates. Substrate thicknesses of 0.015" to 0.500" and greater are available. The base substrate is resistant to all etchants and solvents used in printed circuit production. Consequently, all common PWB processes can be used to produce TMM microwave laminates.

(See reverse for product data)

ROGERS

TMM® Temperature Stable Microwave Laminate

PROPERTY	METHOD	CONDITION	UNITS	DIRECTION	TYPICAL VALUES				
					TMM 3	TMM4	TMM6	TMM10	TMM10i**
Dielectric Constant ε_r	IPC-TM-650 Method 2.5.5.5	10GHz	—	Z	3.27 ±0.016	4.50 ±0.045	6.00 ±0.080	9.20 ±0.230	9.80 ±0.245
Thermal Cofficient of ε_r	IPC-TM-650 Method 2.5.5.5	-55 to + 125°C	ppm/°K	Z	+37	--	-11	-38	-43*
Dissipation Factor	IPC-TM-650 Method 2.5.5.5	10 GHz	—	Z	0.0020	0.0020	0.0023	0.0022	0.0020
Insulation Resistance	ASTM D257	C96/60/95	Gohm		>2000	>2000	>2000	>2000	>2000*
Volume Resistivity			Mohm•cm		3 X 10^9	6 X 10^8	1 X 10^8	2 X 10^7	—
Surface Resistivity			Mohm		>9 X 10^9	1 X 10^9	1 X 10^9	4 X 10^7	—
Copper Peel Strength	IPC-TM-650 Method 2.4.8	After solder float 1 oz. EDC	lb/inch	X,Y	3.0	3.0	3.0	3.0	3.0
Flexural Strength	ASTM D790	A	kpsi	X,Y	16.53	15.91	15.02	13.62	—
Flexural Modulus	ASTM D790	A	Mpsi	X,Y	1.72	1.76	1.75	1.79	1.80*
Impact, Notch Izod	ASTM D256A		ft-lb/in	X,Y	0.33	0.36	0.42	0.43	—
Water Absorption (2X2")	ASTM D570	D48/50	%	—					
3.18mm (0.125") thk					0.06	0.07	0.06	0.09	0.16
1.27mm (0.050") thk					0.12	0.18	0.20	0.20	0.13
Specific Gravity	ASTM D792	A	—	—	1.78	2.07	2.37	2.77	2.77
Specific Heat	Calculated	A	J/g/K		0.87	0.83	0.78	0.74	0.72*
Thermal Conductivity	ASTM F433	25°C	W/m/K	Z	0.68	0.70	0.70	0.73	0.75
Thermal Expansion	ASTM D3386	0 to 140°C	ppm/K	X,Y	16	14	16	16	16*
				Z	20	20	20	20	20*

Notes:
ASTM D3386 corresponds to IPC-TM-650 method 2.4.4.1.
Thermal conductivity and thermal expansion values for TMM-4 and TMM-6 are estimates.
** TMM 10i information is preliminary
*estimated

The information and guidelines contained in this document are intended to assist you in designing with TMM® temperature stable microwave laminates. They are not intended to and do not create any warranties, express or implied, including merchantability or fitness for a particular application. Failure to follow these guidelines may negate any warranties that my otherwise exist. Results may vary as conditions and equipment may vary. The user should determine the suitability of Rogers materials for each specific application.

These products may require a validated export license issued by the United States Department of Commerce for export of these materials from the United States or Canada.

TMM is a registered trademark for Rogers Corporation for its microwave laminates.

ROGERS

Rogers Corporation
Microwave and Circuit Materials Division, 100 S. Roosevelt Avenue, Chandler, AZ 85226
602 961-1382 FAX: 602 961-4533

Printed in U.S.A

Revised 12/96 9313 126 5 0-ON

UL1.2000

ULTRALAM®
2000 Series

Woven Glass Reinforced Polytetrafluoroethylene (PTFE) Microwave Laminate

ULTRALAM 2000 woven glass reinforced PTFE microwave laminate is designed for high reliability stripline and microstrip circuit applications.

Glass reinforcing fibers are oriented in the X/Y plane of the laminate. This orientation maximizes dimensional stability and minimizes etch shrinkage where circuit feature registration is critical.

The dielectric constant of ULTRALAM 2000 is controlled to ± 0.04 from the nominal, within the range of 2.4 to 2.6. It is uniform within each panel, from panel to panel and dissipation factor extends the useful frequency range into K-band.

ULTRALAM 2000 may be cut, sheared and machined to shape. It has excellent resistance to all solvents and reagents, hot or cold, normally used in etching and plating printed circuits.

Cladding options include ½ to 2 oz./ft² (17 to 68 μm thick), rolled or electrodeposited copper and thick metal ground planes.

When requested copper clad ULTRALAM 2000 can be certified to MIL-S-13949H, Type GX or GT.

(See reverse side for product data)

ROGERS

ULTRALAM® 2000 Series

PROPERTY	TEST METHOD	CONDITIONS	UNITS [1]	DIRECTION	TYPICAL VALUE
Dielectric Constant, ε_r	ASTM D3380, 10 GHz	23°C	—	Z	2.4-2.6 ± 0.04
Dissipation Factor, Tan δ	ASTM D3380, 10 GHz	23°C	—	Z	0.0022 max. 0.0019 typ.
Volume Resistivity	ASTM D257	C96/23/95	Mohm•cm	Z	2.0×10^7
Surface Resistivity	ASTM D257	C96/23/95	Mohm	X,Y	4.1×10^7
Dielectric Breakdown	ASTM D149 (parallel)	D48/50	kV	X,Y	>50
Arc Resistance	FED-STD-406-4011		sec.	X,Y	185
Tensile: Modulus	ASTM D638	A	GPa(kpsi)	X Y	11.7 (1700) 9.0 (1300)
Strength	ASTM D638	A	MPa(kpsi)	X Y	147 (21.3) 136 (19.7)
Compressive: Modulus	ASTM D695	A	GPa (kpsi)	X Y	11.0 (1600) 9.0 (1300)
Strength	ASTM D695	A	MPa (kpsi)	X Y	>70 (>10.2) 58 (8.4)
Flexural Strength	ASTM D790	A	MPa (kpsi)	X Y	170 (24.6) 104 (15.1)
Water Absorption	ASTM D570	D96/50	%	—	0.03
Copper Peel Strength	MIL-P-13949 (1 oz. ED)	A Solder Elev. temp. Proc. Sol.	kNm^{-1} (lb/in) kNm^{-1} (lb/in) kNm^{-1} (lb/in) kNm^{-1} (lb/in)	X,Y X,Y X,Y X,Y	3.25 (18.6) 2.38 (13.6) 3.01 (17.2) 3.13 (17.9)
Coefficient of Thermal Expansion	ASTM E831	25 to 150°C	ppm/°C	X Y Z	9.5 9.5 120

[1] S1 units given first, with other frequently used units in parentheses.

ROGERS

Rogers Corporation
Microwave and Circuit Materials Division, 100 S. Roosevelt Avenue, Chandler, AZ 85226 602 961-1382 FAX: 602 961-4533

Printed in U.S.A. Revised 7/93

Appendix G
Microwave Connectors

Connector Type	Reference to Second Edition
SMA Connector	Figure 7.4
TNC Connector	Figure 7.6
Type N Connector	Figure 7.8
APC-7 Connector	Figure 7.9
K Connector	Figure 7.13

SMA Connector

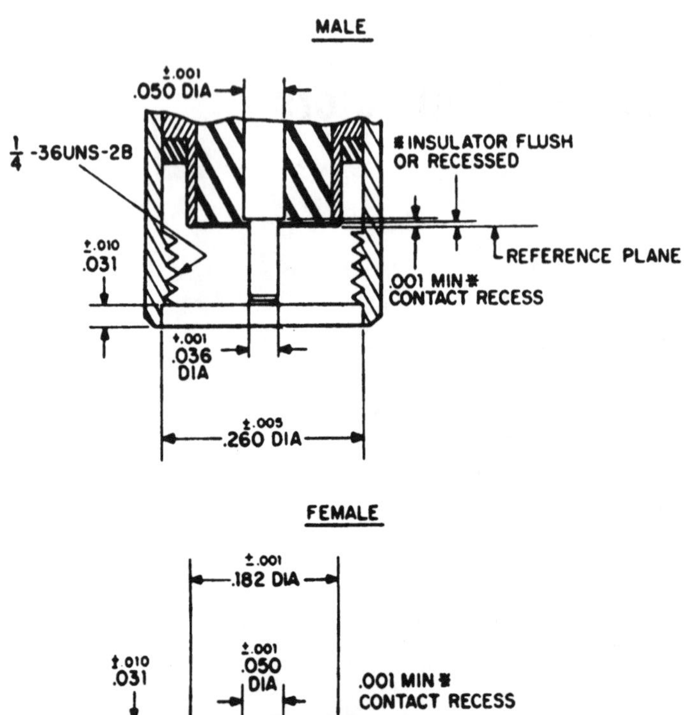

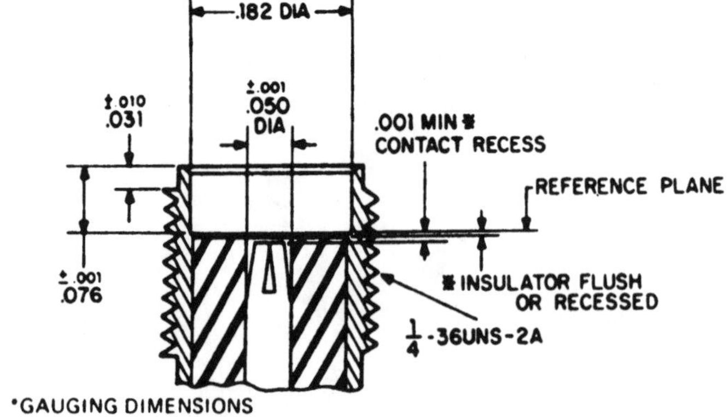

*GAUGING DIMENSIONS

TNC Connector

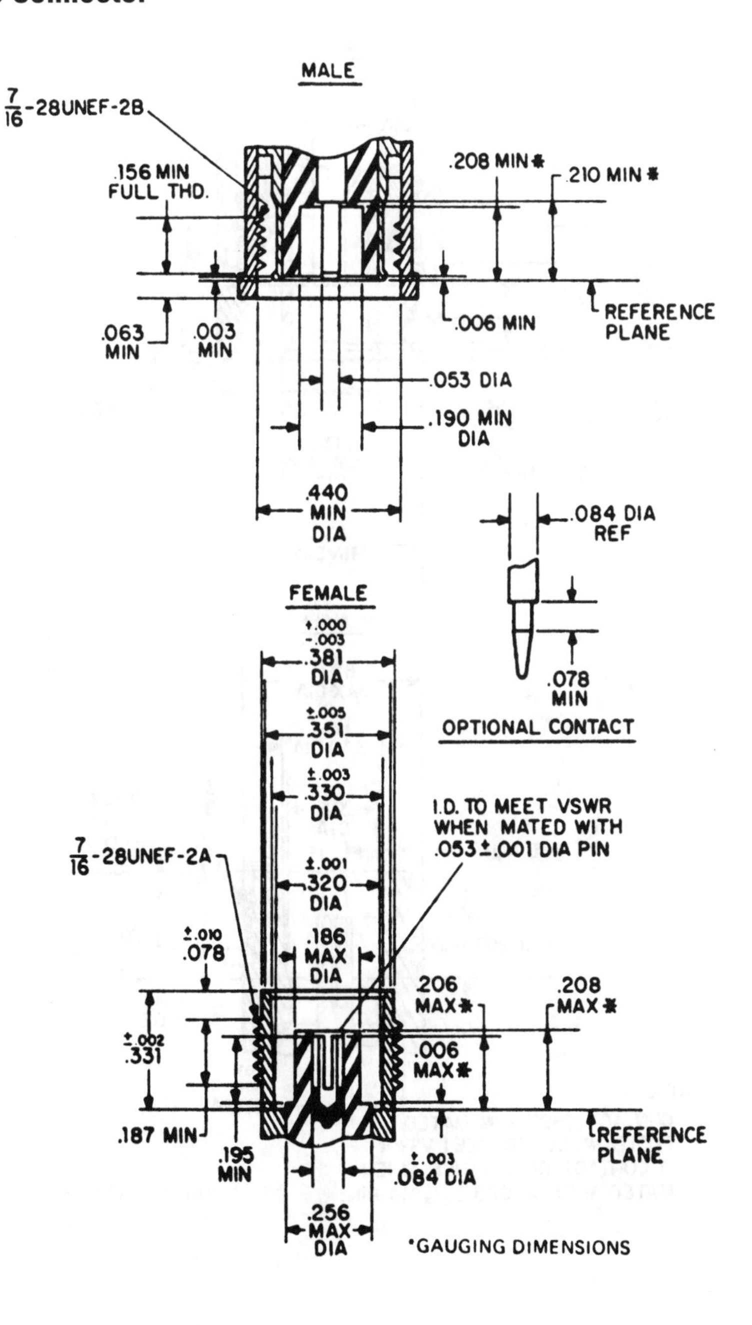

Type-N Connector

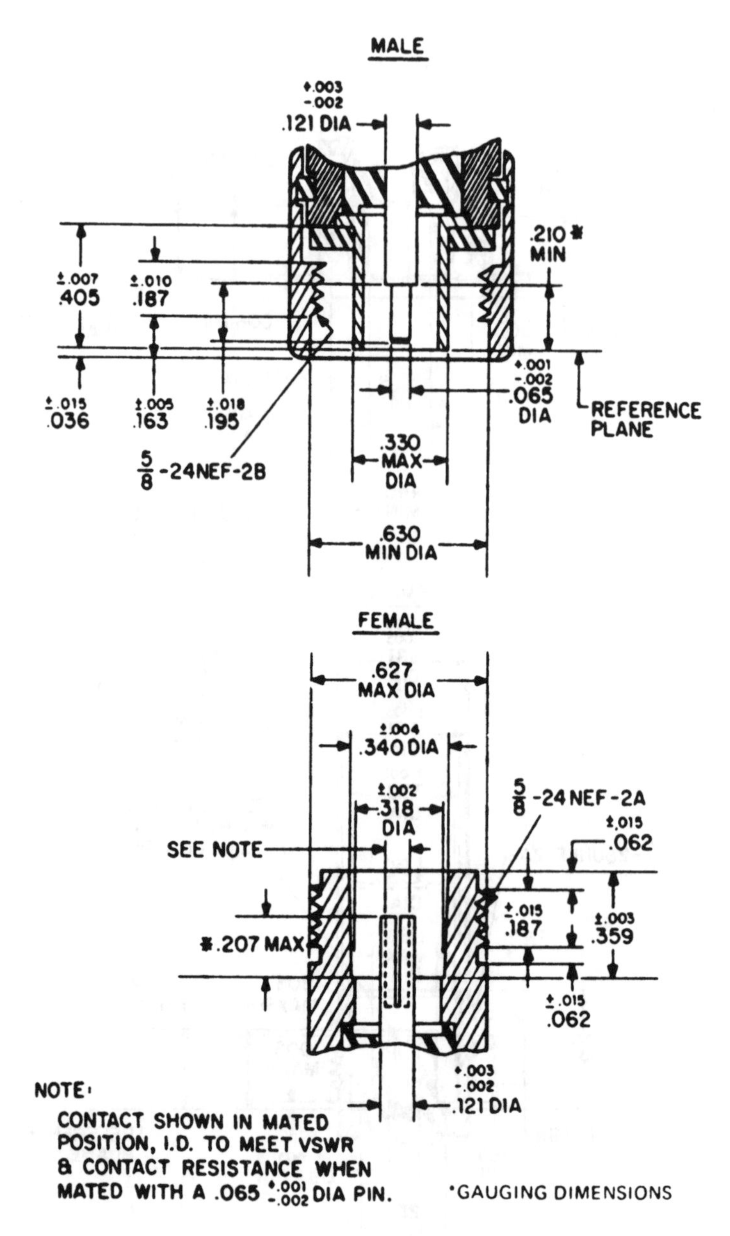

APC-7 Connector

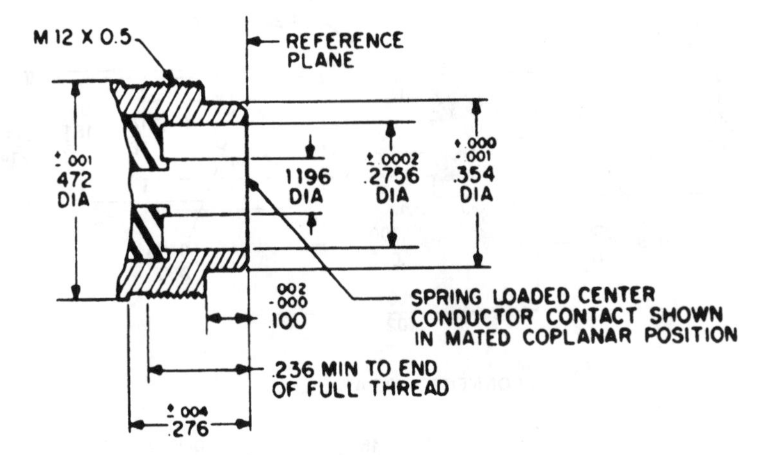

K Connector

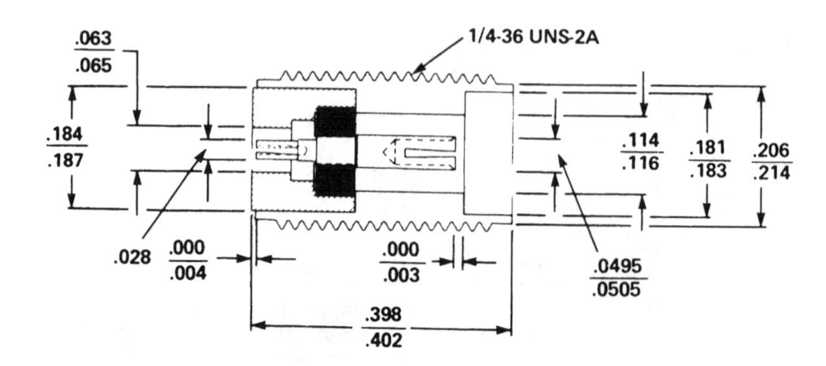

K CONNECTOR (SPARKPLUG)

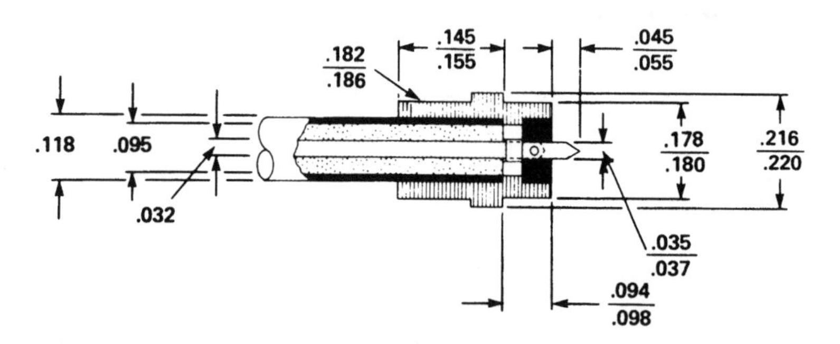

K MALE CONNECTOR AND CABLE
(MALE COUPLING NUT NOT SHOWN)

Bibliography

Chapter 1

Indium Corporation of America, "Indalloy Specialty Solders," Form No. 102-8655 CP10M, Utica, NY, revised, May 1980.

Manko, H. H., *Solders and Soldering,* New York, NY: McGraw-Hill, 1964.

Chapter 2

Borasee, V., "Substrates Influence Thin Film Performance," *Microwaves,* October 1982.

——— "Electrical Properties Govern Substrate Effectiveness," *Microwaves & RF,* February 1983.

Donegan, T., "PTFE Substrates Require Special Care in Fabrication," *Microwaves,* October 1982.

Fogiel, M., *Modern Microelectronic Circuit Design, IC Applications, Fabrication Technology,* Vol. 1, Staff of Research and Education Association, New York, NY, 1981.

Graff, R. F., *Modern Dictionary of Electronics,* 5th Ed., Indianapolis, IN: Howard W. Sams, 1982.

——— *The Basics of Sputtering,* 3rd Ed. Orangeburg, NY.

Nowicki, T. E., "Microwave Substrates Present and Future," in *Microwave Tech Topics,* Vol. 2, 3M Co., St. Paul, MN.

Olyphant, M., "Measuring Anisotrophy in Microwave Systems," *IEEE MTT-S,* 1979.

Olyphant, M., and T. E. Nowicki, "Microwave Substrates Support MIC Technology," Vols. 1 and 2, *Microwaves,* November and December 1980.

Olyphant, M., D. D. Demeny, and T. E. Nowicki, "Epsilam 10—A New High Dielectric Constant Conformable Copper-Clad Laminate," *Cutips,* No. 6, 3M Co., St. Paul, MN.

Rogers Corporation, "The Advantage of Nearly Isotropic Dielectric Constant for RT/Duroid 5870-80 Glass Microfiber-PTFE Composite," Chandler, AZ, 1981.

Reference Data for Radio Engineers, 6th Ed., Indianapolis, IN: Howard W. Sams, 1975.

Rogers Corporation, Soladyne Division, "Design Guide for PTFE-Based Circuit Boards," 1989.

Vossberg, W. A., "Stripping the Mystery from Stripline Laminates," *Microwaves,* January 1968.

Chapter 3

Graff, R. F., *Modern Dictionary of Electronics,* 5th Ed., Indianapolis, IN: Howard W. Sams, 1982.

Harper, C. A., *Handbook of Thick Film Hybrid Microelectronics,* New York, NY: McGraw-Hill, 1974.

Indium Corporation of America, *Understanding Solders and Soldering,* Utica, NY, 1981.

Isacoson, N. I., "Automated Mask Generation: A Price/Performance and Practical Discussion," *MSAT 83,* Washington, DC, March 1983.

Laverghetta, T. S., *Modern Microwave Measurements and Techniques,* Norwood, MA: Artech House, 1988.

Manko, H. H., *Solders and Soldering,* New York, NY: McGraw-Hill, 1964.

Scott, E. C., and F. A. Kanda, *The Nature of Atoms and Molecules,* New York, NY: Harper and Row, 1962.

State of the Art, Inc., *The Fundamentals of Thick Film Hybrid Technology,* Ch. 16, 1970.

Chapter 4

Bergman, D., "Beyond Gerber, The Birth and Growth of GenCAM," *Printed Circuit Design*, April 1999, pp. 10–16.

Nedbal, R., "Zen and the Art of CAD-to-CAM Data Transfer," *Printed Circuit Design*, July 1999, pp. 30–33.

Rogers Corporation, Soladyne Division, "Design Guide for PTFE-Based Circuit Boards," 1989.

Rogers Corporation, Soladyne Division, "Solutions," Solution #8006, 1990.

Chapter 5

Harper, C. A., *Handbook of Thick Film Hybrid Microelectronics,* New York, NY: McGraw-Hill, 1974.

Ramachandra, K., Filtran Microcircuits, "Gold Plating Considerations for Microwave Circuits," 1997.

Reference Data for Radio Engineers, 6th Ed., Indianapolis, IN: Howard W. Sams, 1975.

Rogers Corporation, Soladyne Division, "Design Guide for PTFE-Based Circuit Boards," 1989.

Rogers Corporation, "Through-Hole and Edge Plating of RT/Duroid® Microwave Laminates," RT 4.2.1, 1994.

Standards and Specifications Committee of ISHM, "Glossary of Terms for Hybrid Microelectronic Standards Guidelines," *ISHM STD02,* Montgomery, AL, April 1975.

Tramposch, R., "Thin Film Processing of Hybrid IC's," *The Microwave Systems Designer's Handbook,* Palo Alto, CA: EW Communications, Inc., 1983.

Vossberg, W., "Stripping the Mystery From Stripline Laminates," *Microwaves,* January 1968.

Chapter 6

Emerson & Cumming, "Circuit Assembly Adhesives Selector Guide," 1999.

Fogiel, M., *Modern Microelectronic Circuit Design, IC Applications, Fabrication Technology,* Vol. 1, Staff of Research and Education Association, New York, 1981.

Graff, R., *Modern Dictionary of Electronics,* 5th Ed., Indianapolis, IN: Howard W. Sams, 1982.

Indium Corporation of America, *Understanding Solders and Soldering,* Utica, NY, 1981.

Johnson, D. R., and E. L. Chanez, *Characterization of the Thermosonic Wire Bonding Techniques,* Albuquerque, NM: Sandia Laboratories.

Kester Solder Co., "Problems in Soldering Gold Plate," Laboratory Bulletin, January 1963.

Kulesza, F. W., "New Epoxy Systems for Microelectronics," Billerica, MA: Epoxy Technology, Inc., January 1976.

Manko, H. H., *Solders and Soldering,* New York, NY: McGraw-Hill, 1964.

Small Precision Tools, *Bonding Handbook,* San Rafael, CA: Precision 79, 1977.

Sergent, J., and E. C. Thompson, "Thick and Thin Film Hybrid Microcircuits Professional Advancement Course," Cahners Exposition Group, Boston, MA, 1982.

3M Co., "A User's Guide to Vapor Phase Soldering With Flourimer® Electronic Liquids," St. Paul, MN, 1981.

Thwaites, C. J., *Soft Soldering Handbook,* Middlesex, UK: International Tin Research Institute, 1982.

Yost, F. G., "Soldering to Gold Films," *Gold Bulletin,* Vol. 10, No. 4, October 1977.

Chapter 7

Berson, B., "Microwave Packaging Evolves To Meet Design and Market Challenges," *MSN & CT,* September 1985.

Bilski, S., "Microwave Tutorial," *Hybrid Circuit Technology,* June 1988.

Broekhulzen, F., "Mechanical Design Tips for EMI Shielding," *Applied Microwaves & Wireless,* September 1999, pp. 80–84.

Dietz, R., "Design and Fabrication of MIC's Using Heavy Metal Backed Substrates," *Microwave Hybrid Circuit Conference,* Sedona, AZ, September 1988.

Lakshninarayanan, V., "Basic Steps to Successful EMC Design," *RF Design,* September 1999, pp. 34–48.

March, S., "Microstrip Packaging: Watch the Last Step," *Microwaves,* December 1981.

Markstein, H. W., "Miniaturized Microwave Packaging—A Different Game," *Electronic Packaging and Production,* August 1982.

About the Author

Thomas S. Laverghetta is a professor at Purdue University's Fort Wayne campus. He teaches all the electronic communications courses at Purdue, including an introductory microwave course.

Mr. Laverghetta spent 23 years in the industry prior to teaching full time. He was a design engineer at ITT Aerospace, Magnavox, Anaren Microwave, and General Electric Heavy Military Electronic Systems. Prior to receiving his BSEE from Syracuse University, he was a microwave technician at the Syracuse University Research Corporation and at General Electric.

Mr. Laverghetta has designed stripline and microstrip circuitry for many space applications, shipboard and ground equipment, airborne equipment, and test stations.

He is the author of *Microwave Measurements and Techniques*, *Handbook of Microwave Testing*, *Microwave Materials and Fabrication Techniques*, three editions, *Practical Microwaves*, *Modern Microwave Measurements and Techniques*, *Solid State Microwave Devices*, and *Analog Communications for Technology*. He has also authored many papers for journals and conference proceedings.

Mr. Laverghetta received his MSEE from Purdue University, is a Senior Member of IEEE and a member of ASEE (American Society for Engineering Education), and is an Accredited Professional Consultant through the American Consultants League.

Index

Recent Titles in the Artech House Microwave Library

Advanced Automated Smith Chart Software and User's Manual, Version 3.0, Leonard M. Schwab

Behavioral Modeling of Nonlinear RF and Microwave Devices, Thomas R. Turlington

Computer-Aided Analysis of Nonlinear Microwave Circuits, Paulo J. C. Rodrigues

Design of FET Frequency Multipliers and Harmonic Oscillators, Edmar Camargo

Design of RF and Microwave Amplifiers and Oscillators, Pieter L. D. Abrie

EPFIL: Waveguide E-plane Filter Design, Software and User's Manual, Djuradj Budimir

Feedforward Linear Power Amplifiers, Nick Pothecary

Generalized Filter Design by Computer Optimization, Djuradj Budimir

GSPICE for Windows, SigCad Ltd.

Introduction to Microelectromechanical (MEM) Microwave Systems, Hector J. De Los Santos

Microwave Engineers' Handbook, Two Volumes, Theodore Saad, editor

Microwave Filters, Impedance-Matching Networks, and Coupling Structures, George L. Matthaei, Leo Young, and E.M.T. Jones

Microwave Materials and Fabrication Techniques, Third Edition, Thomas S. Laverghetta

Microwave Mixers, Second Edition, Stephen Maas

Microwave Radio Transmission Design Guide, Trevor Manning

Microwaves and Wireless Simplified, Thomas S. Laverghetta

Neural Networks for RF and Microwave Design, Q. J. Zhang and K. C. Gupta

PACAD: RF Linear Power Amplifier Design, Software and User's Manual, Ramin Fardi, Keyvan Haghighat, and Afshin Fardi

RF Design Guide: Systems, Circuits, and Equations, Peter Vizmuller

The RF and Microwave Circuit Design Handbook, Stephen A. Maas

RF and Microwave Coupled-Line Circuits, Rajesh Mongia, Inder Bahl, and Prakash Bhartia

RF Power Amplifiers for Wireless Communications, Steve C. Cripps

RF Systems, Components, and Circuits Handbook, Ferril Losee

TRAVIS Pro: Transmission Line Visualization Software and User's Manual, Professional Version, Robert G. Kaires and Barton T. Hickman

Understanding Microwave Heating Cavities, Tse V. Chow Ting Chan and Howard C. Reader

For further information on these and other Artech House titles, including previously considered out-of-print books now available through our In-Print-Forever® (IPF®) program, contact:

Artech House
685 Canton Street
Norwood, MA 02062
Phone: 781-769-9750
Fax: 781-769-6334
e-mail: artech@artechhouse.com

Artech House
46 Gillingham Street
London SW1V 1AH UK
Phone: +44 (0)20 7596-8750
Fax: +44 (0)20 7630 0166
e-mail: artech-uk@artechhouse.com

Find us on the World Wide Web at:
www.artechhouse.com